Traffic-Related Air Pollution and Exposure in Urbanized Areas

Traffic-Related Air Pollution and Exposure in Urbanized Areas

Bernard Połednik, Sławomira Dumała, Łukasz Guz and Adam Piotrowicz
Lublin University of Technology, Lublin, Poland

Routledge
Taylor & Francis Group

LONDON AND NEW YORK

First published 2021
by Routledge/Balkema
Schipholweg 107C, 2316 XC Leiden, The Netherlands
e-mail: enquiries@taylorandfrancis.com
www.routledge.com – www.taylorandfrancis.com

Routledge/Balkema is an imprint of the Taylor & Francis Group, an informa business

Library of Congress Cataloging-in-Publication Data
A catalog record has been requested for this book

ISBN: 978-1-032-01281-0 (hbk)
ISBN: 978-1-032-07875-5 (pbk)
ISBN: 978-1-003-20614-9 (ebk)

DOI: 10.1201/9781003206149

Typeset in Times New Roman
by codeMantra

Contents

Preface

Traffic-related pollutions, apart from the emissions from other anthropogenic sources, significantly decrease the air quality in cities. This is especially true in the case of the areas located in the vicinity of communication routes with high traffic intensity. In accordance with epidemiological studies, the transport emissions affecting road users, both drivers and pedestrians, have a detrimental effect on health and may contribute to numerous serious illnesses.

This work presents the characteristics of traffic-related pollution and the factors affecting its concentration in the air. The effects of exposure to traffic-related pollution on health are described as well. The transport emissions in Lublin, Poland, are discussed in detail. The results of previous studies on traffic-related environmental pollution are presented, including pollutant concentrations as well as exposures of drivers and pedestrians along one of the busiest roads. The studies account for different seasons and weather conditions. The methods of reducing traffic-related pollution and prospects for improving the urban air quality are described as well.

This book is intended for academics, researchers, engineers, and professionals involved in studies and assessment of the road transport impact on the environment.

About the authors

Sławomira Dumała, Ph.D. Eng., graduated from the Lublin University of Technology in 2002 and received a Master's degree in Environmental Protection with a specialization in Heating, Ventilation, and Air Protection. Since 2005, she has been working at the Faculty of Environmental Engineering of Lublin University of Technology. She received a Ph.D. degree in Technical Sciences in the discipline of Environmental Engineering. She is a double scholarship holder under the Human Capital Operational Program – "Scholarship for PhD students II" and "Scientific scholarships for PhD students working within research teams". She is authorized to prepare energy certificates and characteristics and is an expert on thermal imaging research. Currently, she is involved in the study of indoor air quality, methods of cleaning, and removing aerosol and bio-aerosol pollutants.

Łukasz Guz, Ph.D. Eng., was born in 1981 in Lubartów, Lublin voivodship, Poland. In 2007, he received an M.Sc. degree at the Faculty of Environmental Engineering of Lublin University of Technology. Since that time, he has been working in Department of Environmental Protection Engineering, and in 2018, he successfully defended his Ph.D thesis. His main fields of scientific activity are indoor and outdoor air pollution, measurement methods in environmental engineering, HVAC, and renewable energy systems as well as building physics. He published over 36 papers, 19 chapters in monographies, 11 Polish patents, and 38 Polish patent applications.

Adam Piotrowicz, Ph.D. Eng., was born in Lublin, Poland. In 2001, he received an M.Sc. degree in Environmental Protection from the Lublin University of Technology. In 2012, he successfully defended his Ph.D. thesis at his alma mater. For many years he has been employed at the Faculty of Environmental Engineering of the Lublin University of Technology, currently as an Assistant Professor. His research interests are related to air pollution from anthropogenic sources in urban areas. He is the author or co-author of several dozen publications in the field of environmental engineering and he has several patents.

Bernard Połednik is an Associate Professor in Environmental Engineering at the Lublin University of Technology, Poland. He received his Master's degree in Physical Chemistry from AGH University of Science and Technology in Cracow, Poland. He completed post-graduate studies in Environmental Engineering and received a Ph.D. degree from the Lublin University of Technology. His professional interests focus on environmental contaminations and the quality of indoor and outdoor air.

Chapter 1

Pollutant characteristics and emissions

For many years, intensive development of the automotive industry has been observed, most noticeably in larger cities and urban agglomerations. In urban areas, where the land cover is conducive to the absorption of solar energy, a specific microclimate is created, characterized by higher temperature and lower humidity. Moreover, dense build-up results in the attenuation of wind strength, which limits the circulation of air masses and, thus, the ability to ventilate. All of this is compounded by a significant traffic volume; thus, road transport, as one of the main sources of air pollution, exerts a very negative impact on the environment. An important feature of road transport pollution is that – compared to sectors such as energy and industry – the amount of exhaust fumes emitted by vehicles is smaller, but the concentration levels of pollutants in the proximity of linear sources such as roadways can be significantly increased. Considering that these are often densely populated areas, the problem is really severe.

Typically, exhaust emissions are taken into account when considering air pollution from vehicles. The exhaust gases comprise carbon monoxide, nitrogen oxides, hydrocarbons, sulfur oxides and particulate matter. Exhaust emissions vary depending on the engine type. This results directly from the use of a different type of fuel but also from a different way of preparing the combustible mixture, and, finally, a different process of combustion itself. The movement of vehicles is also related to non-exhaust emissions, which should also be taken into consideration. These include the abrasion of the top layer of tires, wear of brake and clutch linings, as well as road surface abrasion and resuspension of the road dust.

Various actions are being taken to eliminate the harmful effects of transportation. On the one hand, the aim is to systematically reduce the amount of fuel consumed, while constantly improving the methods of its combustion. More and more modern vehicle constructions using many new pro-ecological solutions are being and have been introduced to the market. Furthermore, propulsion sources alternative to fossil fuels are commonly used. At the same time, however, in many European countries, the number of vehicles is increasing at a high rate and the fleet itself is largely obsolete. Nevertheless, new technological solutions make it necessary to look at the problem of the environmental impact of vehicles more widely. This is, for example, the case with electric cars, commonly referred to as zero emission cars. Although they do not emit fumes while driving, the way in which electricity is generated should be taken into account. Another environmental problem is the method of producing and, later, utilizing the batteries necessary for electricity storage. It should also be noted that all vehicles, regardless of the propulsion system used, are the source of non-exhaust emissions.

DOI: 10.1201/9781003206149-1

One of the most harmful and dangerous components of exhaust gases is carbon monoxide (CO), produced when carbon-containing compounds are incompletely burned. It is a colorless, odorless and tasteless gas that, after entering the human body, reacts with hemoglobin (Hb), leading to the formation of carboxyhemoglobin (COHb). As a result, the process of combining hemoglobin with oxygen is blocked, as the affinity of carbon monoxide for hemoglobin is 200–250 times greater than that of oxygen (Blumenthal 2001, Bleecker 2015, Rose et al. 2020), Depending on the concentration of CO in the air, hypoxia can manifest itself in the form of fatigue, weakness, headaches and dizziness, disorientation, and, in the case of higher concentrations, loss of consciousness and death by suffocation. Lower doses of CO combined with prolonged exposure may lead to damage to the central nervous system and, thus, decrease in visual perception and driving performance (Raub and Benignus 2002, Weaver et al. 2007).

In addition to nitrogen oxides and volatile organic compounds, CO is referred to as a ground-level ozone precursor because in the presence of sunlight, it takes part in a series of chemical reactions leading to the formation of photochemical smog. In vehicles, carbon monoxide is formed when fuel is burned with an insufficient amount of oxygen (partial oxidation). Comparing the CO emissions by engine type, the combustion of fuel in a diesel engine produces less CO due to the fact that diesel engines have a high excess air ratio. Much of the carbon monoxide is retained on the catalytic converter, where CO is converted into CO_2.

Other environmentally important components of exhaust gases are nitrogen oxides (NO_x). Inhalation of NO_x leads to irritation of the throat mucous membranes and can be fatal at higher concentrations. Two compounds belong to this group: nitric oxide (NO) and nitrogen dioxide (NO_2). Nitric oxide is a colorless and odorless gas that is slightly soluble in water. Its molecule is unstable and quickly oxidizes to nitrogen dioxide in the air. Nitrogen dioxide, in turn, is a brown gas that dissolves very well in water and has a pungent, characteristic odor. It is highly toxic and, even after a short-term exposure leads to irritation of eyes and respiratory system, which in turn results in breathing problems. Moreover, NO_2 has the ability to oxidize iron in hemoglobin, as a result of which it loses its oxygen-carrying capacity. In the atmosphere, nitrogen dioxide forms nitric acid that is leached out as nitrates by rainfall.

In the case of road transport, nitrogen oxides are released during the high-temperature combustion of fuels. In spark-ignition engines, where fuel combustion occurs at low air–fuel equivalence ratio (lambda), the formation of nitrogen oxides is limited. The problem is much more acute in diesel vehicles where the combustion process takes place at high temperatures with a high excess air ratio.

Road transport is a major source of nitrogen oxides. In some EU countries, the share of emissions from road transport can exceed 50%. Despite this, emissions have been steadily decreasing over the last few decades, mainly due to new cars being fitted with catalytic converters or catalyst systems. In the case of diesel engines, an SCR technology, i.e. selective catalytic reduction of nitrogen oxides, has been used since the 1980s. Depending on the active catalytic components, vanadium and tungsten-vanadium catalysts are used, but manganese, molybdenum, platinum and palladium catalysts are also possible. With the catalyst, the reduction of NO_x to molecular nitrogen N_2 and water (water vapor) takes place with the aid of ammonia as a reducing agent. For safety reasons, a non-toxic and odorless aqueous urea solution of 32.5% is used. The solution decomposes into ammonia and carbon dioxide when exposed to high flue gas temperatures. In spark-ignition engines, on the other hand, so-called three-way catalysts are used, which enable the reduction of nitrogen oxides, with the simultaneous oxidation of carbon monoxide and hydrocarbons. In the case of the latter, which are also an important component of exhaust gases, a large reduction in their emissions is precisely due to the equipping of new vehicles with catalytic converters.

As a result of the combustion of fuel contaminated with sulfur, sulfur oxides are emitted. Among this mixture of compounds, the main share (about 90%) is sulfur dioxide (SO_2). It is a colorless gas with a pungent and suffocating odor. Inhalation of sulfur oxides leads to irritation of the mucous membrane of the nose, throat and eyes, and even to death at higher concentrations.

In the air, SO_2 is oxidized to SO_3, and this in turn to sulfuric acid. The process takes place with the participation of catalysts in the form of elements such as chromium, aluminum, vanadium, or manganese adsorbed on particulates. Sulfur dioxide is a well-known compound that contributes to the formation of acid rain. Its presence in the atmosphere causes a number of negative effects, such as corrosion of buildings. In combination with nitrogen oxides, carbon oxides and particulates, it is responsible for the formation of London smog.

Over the years, as a result of stringent regulatory requirements, the sulfur content of fuel has been significantly reduced. This applies to both spark and compression ignition engines. As a consequence, sulfur oxides originating from motorization have a small share in total emissions and, therefore, do not constitute a significant environmental problem at present.

The use of fuels in motor vehicles leads to the formation of particulate matter (PM). PM has a liquid or solid state and includes elemental and organic carbon, nitrogen and sulfur compounds, hydrocarbons, as well as heavy metals. Metals can be released through the abrasion of engine parts, as fuel additives or as contaminants. For many years, there has been an increasing proportion of PM emissions from non-exhaust sources such as tire wear and brake and clutch lining wear.

One of the main criteria for classifying PM is size. Currently, the most commonly used classification includes total suspended particles (TSP), coarse particles (PM_{10}), fine particles ($PM_{2.5}$) and ultrafine particles. Definitions of PM_{10} and $PM_{2.5}$ can be found in the Directive 2008/50/EC of the European Parliament and of the Council of 21 May 2008 on ambient air quality and cleaner air for Europe. PM_{10} is defined as a particulate matter that passes through a size-selective inlet according to the European Standard EN 12341 (current version EN 12341:2014), with a 50% efficiency cut-off at $10\,\mu m$ aerodynamic diameter. On a similar basis, $PM_{2.5}$ is defined as a particulate matter that passes through a size-selective inlet according to the European Standard EN 14907 (currently superseded by EN 12341:2014), with a 50% efficiency cut-off at $2.5\,\mu m$ aerodynamic diameter (EC 2008).

In general, the formation of PM is accompanied by complex processes of a physicochemical nature. It is commonly accepted that diesel engines are more affected by this problem. Due to the lack of air in the combustion chamber, products of incomplete combustion are formed. The soot produced promotes the adsorption of hydrocarbons, nitrogen oxides and sulfur dioxide. Nowadays, in order to reduce particulate emissions, diesel particulate filters (DPFs) are commonly used. For spark-ignition engines, particulate emissions are less of a concern. Only a small amount of soot is formed due to the slower thermal decomposition of gasoline. However, an important source of

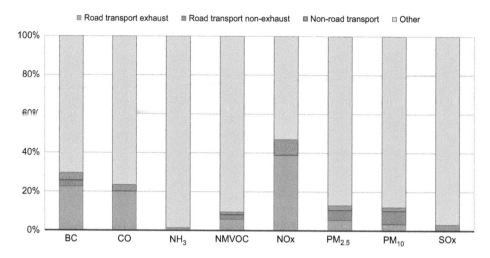

Figure 1.1 Contribution of the transport sector to total emissions of the BC, carbon oxide (CO), ammonia (NH₃), non-methane volatile organic compounds (NMVOC), nitrogen oxides (NOₓ), PM₂.₅, PM₁₀ and sulfur oxides (SOₓ) from road transport in 2018 for EU28. (Based on data reported to the LRTAP Convention provided by EEA, www.eea.europa.eu.)

particulate emissions in this type of engine is engine oil and the contaminants found in gasoline.

In European countries, emissions of major air pollutants from road transport place this sector among significant sources of pollution. Figure 1.1 presents the shares of emissions of some transport-related pollutants compared to total emissions in 2018 considering the 28 EU member states. The data was taken from EU emission inventory database on national emissions reported under the Convention on Long-Range Transboundary Air Pollution (LRTAP Convention). The "non-road transport" sector includes, inter alia, international and domestic aviation, railways, as well as international and domestic shipping. For all but one pollutant, road transport is the main source of emission. The situation is different only for SOₓ, for which international shipping is the dominant source. The percentage share of NOₓ emissions from road transport exhaust sources for all 28 EU countries reached 39% in 2018. During the same year NOₓ emissions from other transport sources amounted to 8%. A high share of emissions from road transport was also observed for black carbon, almost 26%, of which more than 22% represented exhaust emissions. An equally high share was registered for carbon monoxide, over 20%, with the emissions from non-transport sources accounting for only 3%. In addition, transportation by all modes was a notable source of emissions of non-methane volatile organic compounds (over 9%) as well as PM₂.₅ and PM₁₀, 13% and 12%, respectively.

In Europe, for several decades there has been a gradual reduction in emissions of the main air pollutants from road transport, with the simultaneous dynamic development of this sector. According to 2018 data, taking 2000 as a base year, the emissions of nitrogen oxides, carbon monoxide and non-methane volatile organic compounds decreased by 51%, 78% and 81%, respectively. For particulate matter pollutants PM_{10}, $PM_{2.5}$ and black carbon, emissions dropped by 44%, 56% and 66%, respectively.

Reduction of pollutant emissions from road transport is achieved as a result of the introduction of many actions at both international and regional or even local level. On the one hand, EU legislation imposes, e.g. limit values for the concentrations of main pollutants; on the other hand, downtown tolls or low emission zones are introduced in selected places. Other measures aimed at reducing emissions from road transport are associated with the requirements for the quality of fuel sold or developing of Euro emission standards.

In Figure 1.2 are presented the road and non-road transport emissions of NO_x in comparison with the emissions from other sources reported by 28 EU countries in 2000–2018. In addition, the percentage of emissions from road transport in relation to total emissions is also included. In the considered period, both NO_x emissions from road transport and total emissions decreased steadily. In 2005, 2010 and 2018, road transport emissions amounted to 5,027, 3,899 and 2,835 kt, respectively, while total emissions reached 12,274, 9,652 and 7,287 kt, respectively. The shares of emissions from road transport also exhibit a decreasing trend. In the last two years, the values remained below 40%, while in 2005, they amounted to 41% and in 2000 to 43%.

PM emissions, in contrast to nitrogen oxides, include not only exhaust emissions from fuel combustion but also non-exhaust emissions. With the introduction of various solutions in vehicles to reduce the particulate content of exhaust gases, non-exhaust emissions are beginning to play an increasingly prominent role. Figure 1.3 depicts the road transport emissions of $PM_{2.5}$, PM_{10} and TSP reported by the EU28 for the years

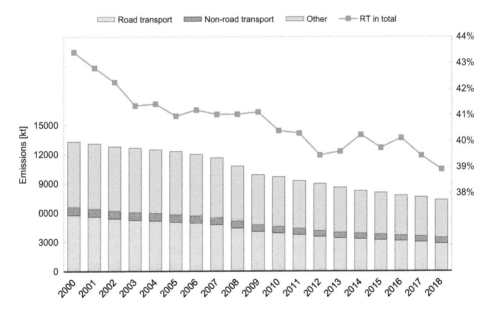

Figure 1.2 NO_x emissions from road transport compared to total emissions for EU28 in the years 2000–2018. (Based on data reported to the LRTAP Convention provided by EEA, www.eea.europa.eu.)

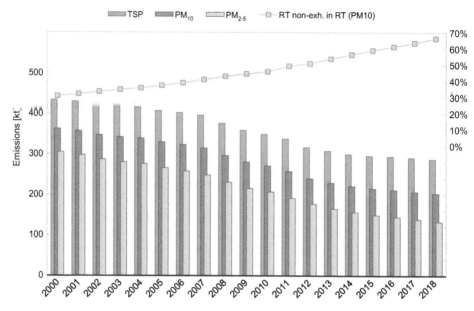

Figure 1.3 $PM_{2.5}$, PM_{10} and TSP emissions from road transport for EU28 in the years 2000–2018. (Based on data reported to the LRTAP Convention provided by EEA, www.eea.europa.eu.)

2000–2018. For all pollutants, there was a clear decrease over the period considered. In 2018, road transport emissions for $PM_{2.5}$, PM_{10} and TSP amounted to 133, 202 and 287 kt, respectively. In 2010, it accounted for 207, 271 and 349 kt, respectively, and in 2000 305, 362 and 433 kt, respectively. Between 2000 and 2018, the share of non-exhaust emissions for PM_{10} increased from less than 31% to more than 66%. With a high degree of probability, further increases can be expected in the future.

In Poland, the national emission inventory is prepared and published annually by the National Centre for Emissions Management (KOBiZE). In the inventory report available for 2018, a new format according to the Nomenclature for Reporting (NFR) was used. The emission estimates have been divided into several sectors, which in turn comprise different categories and subcategories. Road transportation falls into the Energy sector and Transport category, in which aviation, railways, navigation, and other transportation are also contained. In the subcategories of road transportation, the emissions from many different vehicle types are covered, including passenger cars, light commercial trucks, heavy-duty vehicles, and buses as well as mopeds and motorcycles. Moreover, separate subcategories are assigned to gasoline evaporation, automobile tire and brake wear, and automobile road abrasion (EMEP/EEA 2019). The pollutant emissions originating from road transport were calculated with the use of the COPERT 5 software. When estimating emissions, such factors as fuel consumption and characteristics, number of vehicles by category, the mileage by vehicle category and mileage share by road class, the average speed by vehicle category as well as by road class and monthly temperatures were taken into account.

Table 1.1 Emissions of carbon monoxide (CO), nitrogen oxides (NO$_x$ as NO$_2$), sulfur dioxide (SO$_2$), non-methane volatile organic compounds (NMVOC) and ammonia (NH$_3$) from road transportation in 2018, based on the data of the National Centre for Emissions Management (KOBiZE 2020b)

Pollutant	CO	NO$_x$	SO$_2$	NMVOC	NH$_3$
			kt		
Transport	528.30	295.32	0.73	74.90	4.38
Road	526.05	286.74	0.57 (0.1%)	73.96	4.38 (1.4%)
transportation	(22.5%)	(37.6%)		(10.1%)	
Total	2,339.07	761.71	501.93	732.69	316.93

The emissions of the main air pollutants associated with transport in Poland in 2018 are listed in Table 1.1. The values given relate to total emissions from all sectors, emissions from transport including all modes of transport as well as road transport emissions alone, supplemented with the percentage of emissions from road transport in total emissions.

For virtually all of the pollutants presented in the table, transport emissions were mostly made up by road transport. As reported by Statistics Poland, the energy consumption for the needs of individual means of transport has not changed over the past years, and in 2018, it amounted to 93.8% for road transport, 4.5% for air transport, 1.7% for railways, and trace amounts for navigation. Compared to 2008, road transport fuel consumption increased by 46.4% in 2018, with an annual growth rate of 3.9% over the period (Statistics Poland, 2020b).

In 2018, road transport was the major source of nitrogen oxides emissions (37.6%), with a total emission of 761.71 kt. Taking individual subcategories into account, the shares are as follows: heavy-duty vehicles including buses 55.7%, passenger cars 30.6%, light-duty vehicles 13.5%, and mopeds and motorcycles 0.2%. In terms of previous years, NO$_x$ emissions in 2010 represented 33.0% of 881.23 kt and in 2005 27.3% of 867.87 kt. Although total nitrogen oxide emissions have decreased since the early 1990s (from about 1,075 kt in 1990), the emissions from road transport are increasing and have been the main source of these pollutants for many years. The main reason is the significant increase in the number of vehicles and the associated increase in liquid fuel consumption. The great number of older models and the poor condition of a large share of vehicles in use are also of importance.

In Figure 1.4, the NO$_x$ emissions from road transportation in Poland over the period of 2000–2018 including different vehicle categories are presented. After lower emission values between 2012 and 2016, in the last two years the NO$_x$ emissions from road transport have returned to the levels reported for 2007–2011. It can also be seen that compared to the beginning of the analyzed period, there has been a change in the dominant vehicle category. Initially, passenger cars had the largest share, but since 2004, and to a greater extent since 2005, heavy-duty vehicles and buses have become the category with the highest emissions.

As far as other European Union countries are concerned, the highest percentage of the nitrogen oxide emissions from road transport in total NO$_x$ emissions in 2018 was reported for Luxembourg (64.1%), Malta (58.0%), France (56.3%) and Austria (53.5%).

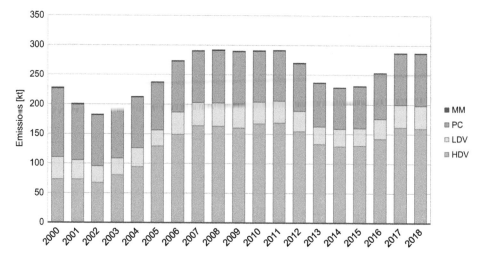

Figure 1.4 Nitrogen oxides (NO$_x$) emissions (expressed as NO$_2$) from road transport in the years 2000–2018 depending on vehicle category: heavy-duty vehicles and buses (HDV), light-duty vehicles (LDV), passenger cars (PC), and mopeds and motorcycles (MM). (Based on data of the National Centre for Emissions Management (KOBiZE, 2020b).)

At the other end of the list, the lowest percentages were recorded in Estonia (22.4%), Finland (24.0%), Greece (28.2%) and Denmark (28.9%). A year earlier, the situation was very similar at both ends of the list. Overall, the share of NO$_x$ emissions in 24 countries exceeded 30%, of which in 12 countries it exceeded 40%. The contributions of each subcategory in 2018 were as follows: passenger cars 49%, heavy-duty vehicles including buses 31%, light-duty vehicles 19%, and mopeds and motorcycles 1%.

According to Directive (EU) 2016/2284 of the European Parliament and of the Council *on the reduction of national emissions of certain atmospheric pollutants*, Poland, like the 27 other European Union countries at the time, is obliged to meet the emission reduction commitments (EC, 2016). Several pollutants are listed in the Directive, namely nitrogen oxides, non-methane volatile organic compounds, sulfur dioxide, ammonia and PM$_{2.5}$. In the case of nitrogen oxides, using 2005 as a base, the reduction should have reached 30% in 2020 and should attain 39% in 2030. Currently, the reduction in NO$_x$ emissions does not exceed 15% (KOBiZE, 2020a). Compared to the rest of Europe, countries such as Denmark, the UK and France have committed themselves to achieving the highest reductions. For the period of 2020–2029, the reduction should reach 56%, 55% and 50%, respectively. In turn, for 2030 and later, the reduction in these countries should amount to 68%, 73% and 69%, respectively. The aforementioned reduction rates for Poland are among the lowest; only Estonia has lower ones – 18% and 30%, respectively. As of 2018, the reduction level commitments for 2020–2029 were not fulfilled by 12 out of 28 countries, apart from Poland, also by Lithuania, Romania, Cyprus and Germany (EEA 2020b). However, for the rest of the EU Member States, the differences between the 2018 NO$_x$ emissions and the reduction commitment for 2020–2029 are not so significant. On the other hand, when comparing the emissions in 2018 with the reduction commitments for 2030 and later, it could be noticed that so far they have not been complied with by any of the countries.

In 2018, road transport in Poland constituted also a significant source of carbon monoxide (22.5%), with a total emission of 2,339.07 kt, ranking second behind household and other small stationary fuel combustion sources (nearly 64%). The largest share was accounted for by passenger cars (65.8%) followed by light-duty vehicles (21.0%), heavy-duty vehicles and buses (7.7%), and mopeds and motorcycles (5.6%). In 2010, the share of CO emissions from road transport amounted to 23.8% of 2,999.11 kt, while in 2005, it reached 26.8% of 2,967.72 kt.

The carbon monoxide emissions from road transport in Poland over the period 2000–2018 depending on vehicle category are presented in Figure 1.5. As for the long-term trend, a systematic decrease in the CO emissions from road transport since 1990 can be observed. This tendency is typical for the vast majority of EU countries. In 2018, the total annual CO emissions for the 28 EU countries amounted to 19,433 kt, of which the road transport emissions accounted for 20.1%. In comparison, in 2010, the total CO emissions equaled 25,924 kt, with a road transport share of 26.6%. In turn, in 2005 the total emission of CO was at the level of 30,653 kt, while the percentage of road transport amounted to 38.2%. According to the data for 2018, the share of CO emissions exceeded 20% in 12 countries, of which it exceeded 40% in six countries. Considering the

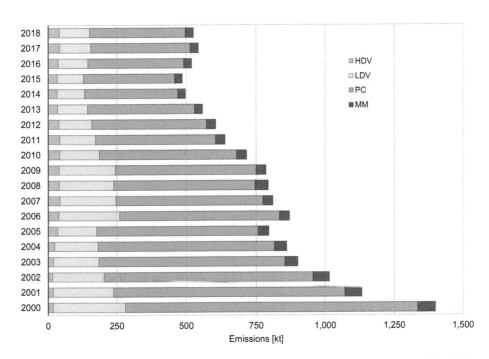

Figure 1.5 Carbon monoxide (CO) emissions from road transport in the years 2000–2018 by vehicle category: heavy-duty vehicles and buses (HDV), light-duty vehicles (LDV), passenger cars (PC), and mopeds and motorcycles (MM). (Based on data of the National Centre for Emissions Management (KOBiZE, 2020b).)

vehicle type, the shares of CO emissions were as follows: passenger cars 70%, mopeds and motorcycles 14%, light-duty vehicles 9%, and heavy-duty vehicles and buses 7%.

Regarding non-methane volatile organic compounds, the emissions from road transport in Poland in 2018 accounted for 10.1% of the total of emissions released (732.69 kt), of which passenger cars were 47.3%; gasoline evaporation, 23.9%; light-duty vehicles, 12.7%; heavy-duty vehicles and buses, 10.6%; and mopeds and motorcycles, 5.5%. The largest share was reported for industrial processes, in particular the *Other solvent and product use* subcategory (over 30%). In the past, the shares of NMVOC emissions from road transport were similar to or higher than that in 2018, for example, reaching 12.3% of 806.45 kt in 2010 and 14.0% of 811.16 kt in 2005. Considering other EU countries, there are significant differences between the percentages of NMVOC emissions from road transport. The largest shares were recorded in Malta (48.7%), followed by Greece (27.4%) and Cyprus (22.5%). In contrast, the lowest contributions were found in Ireland (3.5%), Spain (3.5%) and the UK (3.7%). The ranking has been comparable for the last few years. The percentage share of NMVOC emissions for the 28 EU Member States amounted to 8.2% in 2018 with a total emission of 7,008 kt. In the past it was noticeably higher, reaching 12.0% of 8,260 kt in 2010 and 18.1% of 9,952 kt in 2005. In 2018, the share of NMVOC emissions in eight countries was greater than 10%, in 3 of them it was greater than 20%. The percentages of 2018 NMVOC emissions by vehicle type are as follows: passenger cars, 39%; gasoline evaporation, 29%; mopeds and motorcycles, 20%; heavy-duty vehicles and buses, 6%; and light-duty vehicles, 6%.

NMVOC is another pollutant covered by the emission reduction commitment included in Directive (EU) 2016/2284. In the case of Poland, the reduction of NMVOC emission in relation to 2005 should amount to 25% in 2020 and 26% in 2030 (EC, 2016). In 2018, the emission was reduced to less than 10% or less than 12%, depending on the number of emission sources taken into account (KOBiZE 2020a). However, at the same time, the NMVOC emissions from road transport decreased by almost 35%, mainly due to the changes in the passenger car sector. Among the European Union countries, the largest reduction obligations were assumed by Greece (54% in 2020–2029 and 62% from 2030), Cyprus (45% and 50%, respectively) and France (43% and 52%, respectively). In the case of Hungary, the reduction commitment for 2020–2029 reaches 30%, but as much as 58% thereafter. On the other hand, the lowest level of reduction should be met by such countries as the Netherlands (8% and 15%, respectively), Estonia (10% and 28%, respectively) and Germany (13% and 28%, respectively). According to the NMVOC emissions data for 2018, emission reduction commitments for the period 2020–2029 would not be met by seven countries, in addition to Poland, mainly by Ireland and Malta (EEA 2020b). In turn, the reduction commitments for 2030 and later would be fulfilled in the case of nine countries, headed by Slovakia, Belgium, the Netherlands and Estonia.

In 2018, road transport in Poland was also a minor source of sulfur dioxide (0.1%) and ammonia (1.4%) emissions, in both cases mainly due to passenger cars (61.7% and 89.2%, respectively). As regards SO_2, the coal-based energy industry accounts for the largest share of emissions. However, as a result of stricter emission standards forcing the construction of highly efficient flue gas desulfurization systems, the total emissions have significantly decreased over the years. As for NH_3, the main sources are in the agricultural sector, namely agricultural management (nearly 60%) and manure management (over 35%). In the 28 countries of the European Union, the total SO_2 emissions from road transport accounted for 0.28% of the total emissions in 2018 (0.18% in 2010 and 0.33%

Table 1.2 Total suspended particles (TSP), PM_{10}, $PM_{2.5}$ and BC emissions from road transportation in Poland in 2018, based on data of the National Centre for Emissions Management (KOBiZE 2020b)

Pollutant	TSP	PM_{10}	$PM_{2.5}$	BC
			kt	
Transport	23.30	18.18	13.69	5.89
Road transportation	23.11 (6.1%)	18.00 (7.4%)	13.53 (9.9%)	5.80 (36.5%)
Total	377.70	242.76	136.73	15.91

in 2005). For ammonia, the share in 2018 amounted to 1.3%, while in 2010 and 2005, it reached 2.0% and 2.7%, respectively. For both pollutants, passenger cars accounted for the largest share of emissions, achieving 65% for SO_2 and 90% for NH_3 in 2018.

Road transport is inherently connected with the emission of PM pollutants. Table 1.2 provides an overview of the PM emissions from road transport in comparison to the total inventory emissions for Poland in 2018. The estimates include four pollutants: $PM_{2.5}$, PM_{10}, total suspended particles (TSP) and black carbon (BC).

In Poland, the largest source of particulate matter emissions is small stationary combustion of fuels, mainly in the household sector, ranging from 33% for BC to 49% for $PM_{2.5}$. However, the contribution of mobile sources is also significant. Compared to all modes of transport incorporated in the *Transport* category, road transportation is the predominant source of PM emissions. Of the four pollutants, the highest share was reported for BC (36.5%) with a total emission of 15.91 kt. The shares of BC emissions in 2010 and 2005 amounted to 37.9% of 17.73 kt and 29.4% of 16.71 kt, respectively. Between 2010 and 2015, there was a systematic decrease in BC emissions, followed by a slight increase due to higher fuel consumption. Compared to the previous year, the BC emissions from road transport in 2018 remained at a similar level. According to the data from 2018, emissions of BC from road transportation were mainly caused by passenger cars (42.7%), heavy-duty vehicles and buses (36.7%) and light-duty vehicles (20.4%).

In 2018, the percentage share of BC emissions from road transport for 26 EU countries accounted for 25.7% with a total emission of 194.9 kt. This share has been systematically decreasing, amounting to 37.0% of 285.0 kt in 2010 and 41.2% of 327.3 kt in 2005. Considering the vehicle category, the shares in 2018 were as follows: passenger cars, 46%; light-duty vehicles, 22%; heavy-duty vehicles including buses, 19%; non-exhaust emission, 12%; and mopeds and motorcycles, 1%. On a country basis, Malta had the highest proportion of BC emission (59.6%), followed by Germany (47.4%), Cyprus (44.7%), France (40.9%) and Ireland (40.4%). The lowest percentages were reported for Spain (10.0%), Latvia (10.5%) and Estonia (10.8%). These countries have recorded similar shares in previous years. On the whole, out of 26 countries, the share of BC emissions in 2018 exceeded 20% in 16 countries and 40% in 5 countries.

Among the remaining three particulate matter pollutants, in Poland, according to 2018 data, the highest share of road transport emissions in relation to the total emissions was reported for $PM_{2.5}$ (9.9% of 136.73 kt). For PM_{10} and TSP, this share was slightly lower and amounted to 7.4% of 242.76 kt and 6.1% of 377.70 kt, respectively. In the case of these fractions, sources such as industrial processes or agriculture accounted for a greater share in emissions. Looking back, in 2010 and 2005, these contributions varied as follows: for $PM_{2.5}$ 9.2% of 152.53 kt and 7.2% of 153.63 kt, respectively, for PM_{10} 6.3% of 273.54 kt and 4.8% of 278.27 kt, respectively, whereas for TSP 4.9% of 428.99 kt and 3.7% of 431.36 kt, respectively. Despite the fluctuations in value in the past, an upward trend can generally be observed. Among the considered sources of $PM_{2.5}$ emissions related to road transport, in 2018 heavy-duty vehicles and buses accounted for 26.0%, tire and brake wear for 24.9%, passenger cars for 23.9%, road abrasion for 12.8%, light-duty vehicles for 11.7%, and mopeds and motorcycles for 0.7%. Regarding PM_{10} and TSP, a greater share of non-exhaust sources was estimated, which in the case of $PM_{2.5}$ totaled 37.7%, and for PM_{10} and TSP amounted to as much as 53.2% and 63.5%, respectively.

The methodology adopted for estimating exhaust emissions assumes that particulate matter mass emission factors are related to $PM_{2.5}$ as larger fractions are present in the vehicle exhausts at a negligible level. As a result, the emissions of TSP, PM_{10} and $PM_{2.5}$ are the same for individual vehicle classes. However, as indicated earlier, the differences in PM emissions occur in the case of tire and brake pad wear and road surface abrasion where, unlike in the case of exhaust emissions, large particles predominate. As a result, the estimated non-exhaust emission of PM_{10} is almost two times higher compared to $PM_{2.5}$, and TSP emission is almost three times as high. The emissions of PM pollutants from road transportation in Poland in 2018 together with the shares of individual sources are shown in Figure 1.6.

Considering the situation of 28 EU countries, a general decrease can be observed in the share of $PM_{2.5}$, PM_{10} and TSP emissions from road transport. In 2018, total $PM_{2.5}$ emissions accounted for 10.6% of 1,255 kt, compared to 13.3% of 1,558 kt in 2010 and 15.8% of 1,677 kt in 2005. For PM_{10}, the emissions were 10.2% of 1,989 kt in 2018, 11.5% of 2,364 kt in 2010 and 12.6% of 2,602 kt in 2005. In the case of TSP, the emissions in 2018 amounted to 7.7% of 3,748 kt, in 2010 8.1% of 4,288 kt, and in 2005 8.4% of 4,816 kt. In terms of individual EU countries, the highest percentage of $PM_{2.5}$ emissions

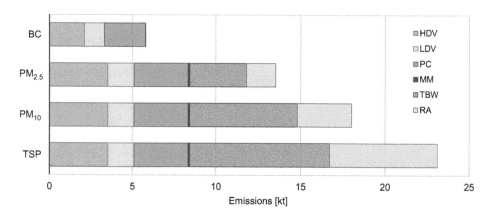

Figure 1.6 Particulate matter emissions from road transport in 2018 including exhaust and non-exhaust sources: Heavy-duty vehicles and buses (HDV), light-duty vehicles (LDV), passenger cars (PC), mopeds and motorcycles (MM), tire and brake wear (TBW) and road abrasion (RA). (Based on data of the National Centre for Emissions Management (KOBiZE, 2020b).)

from road transport in 2018 was recorded by small countries such as Malta (55.7%), followed by Luxembourg (26.7%) and Cyprus (24.3%). In turn, the lowest shares were estimated for Latvia (2.8%), Romania (4.1%) and Croatia (4.7%). In the case of PM_{10}, the largest shares belonged to Malta (45.4%), Sweden (45.3%) and Luxembourg (29.5%), and the lowest to Latvia (2.8%), Romania (4.0%) and Estonia (4.6%). As for TSP, the highest emission shares were reported for Sweden (56.5%), Malta (38.1%) and Luxembourg (31.2%), while the lowest were for Latvia (2.1%), Romania (3.2%) and Denmark (3.5%). Taking into account the shares of individual categories in $PM_{2.5}$ emissions, in 2018 the results were as follows: non-exhaust sources 49%, passenger cars 26%, heavy-duty vehicles and buses 12%, light-duty vehicles 11%, and mopeds and motorcycles 2%. There has been an increase in the percentage share of non-exhaust sources compared to previous years. According to the data for EU countries, in nine of them this share exceeded 50% and in five – even 60%. A similar trend is found for PM_{10} and TSP, where the total contribution of non-exhaust sources is even higher and in 2018 represented 66% and 76%, respectively. As regards PM_{10}, the share exceeded 50% in 22 Member States, 60% in 13, and 70% in 7. In the case of TSP, the share of non-exhaust sources was greater than 50% in 26 countries, 60% in 22, and 70% in 13.

The changes in estimated annual PM emissions from road transport are closely correlated with the changes in fuel consumption in a given year. Comparing the PM emissions in Poland in 2018 with previous years, a clear increase can be observed in relation to 2015 and a comparable level as in 2010, especially in the case of $PM_{2.5}$ and PM_{10}. Emissions of these pollutants from road transportation estimated for the years from 2000 to 2018 including individual emission sources are depicted in Figure 1.7. As for $PM_{2.5}$, it is another pollutant listed in Directive (EU) 2016/2284 in terms of the emission reduction commitments. Starting from 2020, Poland is obliged to reduce the $PM_{2.5}$ emissions by 16% and from 2030 by 58% compared to the base year 2005 (EC, 2016). In 2018, the reduction amounted to 11% (KOBiZE 2020a), although at the same time there was an increase in the $PM_{2.5}$ emissions from road transport by almost 23%.

Within the European Union countries, Cyprus, the Netherlands and Slovakia have taken the largest $PM_{2.5}$ reduction commitments for the period of 2020–2029, with 46%, 37% and 36% respectively. However, from 2030 onwards, these countries have committed themselves to reduce emissions by 70%, 45% and 49%, respectively. In the case of the Czech Republic and Slovenia, the commitments for the 2020–2029 period are 17% and 25%, respectively, but from 2030 and later in both cases they reach as much as 60%. In contrast, the lowest reduction rates have been adopted by Italy (10% in 2020–2029 and 40% from 2030) and Hungary (13% and 55%, respectively). On the basis of the $PM_{2.5}$ emissions data for 2018, the reduction level commitments for the period of 2020–2029 would be met by 17 Member States, and for the period from 2030 onwards by 7 Member States (EEA 2020b).

Road transport is also considered as an emission source of certain amounts of heavy metals. It is assumed that the emission takes place both as a result of combustion processes and from non-exhaust sources. Numerous heavy metals are present in fuel and lubricant oil, namely chromium, copper, zinc, nickel, mercury, lead and cadmium. However, a significant proportion of the emissions may originate from brake lining wear or from tire wear. In the first case, copper and chromium are regarded as the key tracers, while in the second case, zinc in the form of ZnO and ZnS used as a vulcanization agent has the greatest share (Adamiec et al. 2016). Emission inventories

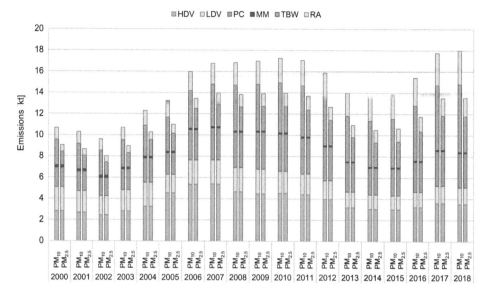

Figure 1.7 PM$_{10}$ and PM$_{2.5}$ emissions from road transport in the years 2000–2018 depending on emission source: heavy-duty vehicles and buses (HDV), light-duty vehicles (LDV), passenger cars (PC), mopeds and motorcycles (MM), tire and brake wear (TBW) and road abrasion (RA). (Based on data of the National Centre for Emissions Management (KOBiZE, 2020b).)

of seven selected heavy metals from road transport in Poland in comparison to total emissions estimated for 2018 are shown in Table 1.3. As with other pollutants, the emissions from road transport are dominant when compared to other modes of transport.

According to the data reported, the largest share in the total copper emission is attributed to road transport, i.e. 34%, in which tribological processes account for the majority. In the case of other metals, the contribution of road transport is not so significant. The main source of chromium emissions were industrial processes, mostly metal industry (28%), while road transport constituted 10%. As far as zinc is concerned, road transport contributed to 6%. The combustion processes of fuels in the manufacturing industry and construction as well as energy industries were indicated as the largest Zn sources with the emission share of 27% and 23%, respectively. For lead and cadmium, as for chromium, the

Table 1.3 Heavy metals emissions from road transportation in 2018, based on data of the National Centre for Emissions Management (KOBiZE 2020b).

Pollutant	Cd	Cr	Cu	Hg	Ni	Pb	Zn
				t			
Transport	0.05	3.54	74.51	0.12	0.56	9.41	27.42
Road	0.04	3.54	74.34	0.12	0.55	9.10	27.41
transportation	(0.4%)	(9.9%)	(34.2%)	(1.4%)	(0.7%)	(3.0%)	(6.2%)
Total	9.45	35.85	217.32	8.74	82.55	303.5	444.75

greatest source of emissions was the metal industry, with a share of 52% and 41%, respectively. In the case of these two metals, the contribution of road transport amounted to 3% for lead and 0.4% for cadmium. Road transport was also a minor source of mercury and nickel emissions, 1.4% and 0.7%, respectively. For both metals, fuel combustion processes in energy industries were identified as the largest source of emissions with a share of 59% for mercury and 37% for nickel. It should be noted that for all the metals mentioned, apart from mercury, the vast majority of emissions originated from non-exhaust sources, i.e. from tire and brake wear. Figure 1.8 shows the changes in emissions of several heavy metals in Poland over the years 2000–2018.

A similar trend in emissions can be observed for copper, chromium and zinc. In 2010, the emissions were around twice higher than in 2000. In 2018, a slight increase was recorded compared to the previous year, but compared to 2010, the emissions increased by about 1.35 times and compared to 2000 more than 2.5 times. Since tire and brake wear is the greatest contributor, the emissions are mostly dependent on vehicle mileage. The situation is different for lead, where the significant decrease in emissions after 2000 is related to the discontinuation of leaded petrol production for road vehicles. However, even in this case, after a decline at the beginning of the century, the estimates of Pb emissions indicate a steady increase.

The heavy metal data reported by the 28 EU countries for 2018 reveals that copper accounted for the largest share of road transport emissions (69%), of which 94% represented non-exhaust emissions and only 4% passenger cars. For zinc and chromium, the emissions accounted for 20% and 17%, respectively, while non-exhaust emissions

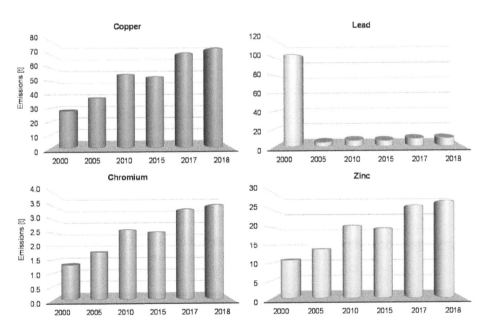

Figure 1.8 Emissions of copper, chromium, zinc and lead from road transport in selected years. (Based on data of the National Centre for Emissions Management (KO-BiZE, 2020b).)

accounted for 79% and 88%, respectively. Lead emissions reached 16%, of which 86% were non-exhaust and 10% passenger cars. Road transport was also a source of mercury at 3% (as exhaust emissions), of which 65% accounted for passenger cars, 23% heavy-duty vehicles, and 11% light-duty vehicles. Considering the changes in emissions in the long term, for Cu, Zn, Cr and Hg a gradual increase can be observed, with a fall between 2008 and 2013. For Pb, the largest reductions in emissions were in the 1990s, with an 88% decrease between 1990 and 2000, while in 2018 compared to 2005 there was a 12% decrease. Taking into account other transport sources, for all metals listed, road transport was the dominant source of emissions.

Vehicles are also taken into account when estimating the emissions of some persistent organic pollutants (POPs). Table 1.4 summarizes the emissions of several POPs caused by road transport in Poland in 2018. In the case of dioxins and furans, the emissions were expressed by means of the International Toxic Equivalent (I-TEQ) which is an index of toxicity relative to the most toxic dioxin – 2,3,7,8-tetrachlorodibenzo-dioxin (TCDD) with the allocated value of 1. The total polycyclic aromatic hydrocarbon (PAH) emission concerns four representatives, i.e. benzo(a)pyrene, benzo(b)fluoranthene, benzo(k)fluoranthene and indeno(1,2,3-cd)pyrene. Dioxins and furans were emitted in greatest parts by extra-industrial combustion processes, as well as during waste management. The dominant sources of PAHs are the extra-industrial combustion processes and production processes (coke production).

In 2018, in Poland, dioxin and furan emissions from road transport accounted for less than 3%, and although they decreased slightly compared to the previous year, an upward trend can generally be observed. This trend is even more noticeable in the case of PAHs, for which emissions from road transport accounted for 0.6%. In contrast, total emissions for both PCDD/PCDFs and PAHs show a rather downward trend. Although the reported shares of road transport in the total POPs emissions are not particularly high, one should be aware that these are highly toxic compounds, which, especially in cities, may have a noticeably negative impact on the environment.

As for EU28 Member States, the total PCDD/PCDF emissions are gradually decreasing. After a period of increase between 2000 and 2008, the emissions from road transport are also decreasing. According to the data for 2018, the share of PCDD/PCDF emission totaled 4%, of which 72% was caused by passenger cars. In the case of PAHs, while total emissions have declined over the years, emissions from road transport have recorded an increase. In 2018, the share of emissions reached 1.5%, of which 58% was accounted for by passenger cars, 25% heavy-duty vehicles and 12% light-duty vehicles.

Table 1.4 Emission of dioxins and furans (PCDD/F), hexachlorobenzene (HCB) and polycyclic aromatic hydrocarbons (PAHs) from road transport in 2018 (KOBiZE 2020b)

Pollutant	Dioxins and furans (PCDD/F)	Hexachlorobenzene (HCB)	Polycyclic aromatic hydrocarbons (PAHs)
	g I-TEQ	kg	t
Total emission	316.07	3.71	231.14
Transport	9.46 (3.0%)	0.01 (0.3%)	1.47 (0.6%)
Road transport	9.32 (2.9%)	0.01 (0.3%)	1.46 (0.6%)

Chapter 2

Overview of road transport equipment in Europe

2.1 CURRENT CONDITION OF VEHICLES AND INFRASTRUCTURE

The data on road infrastructure and means of transport in Europe are available in the Eurostat database, in the subsection: European Commission/Eurostat/Transport: road transport infrastructure, road transport equipment – stock of vehicles (Eurostat 2020).

According to the available statistical data (Eurostat 2020), the total number of passenger cars in Europe in 2018 was 283,285,084, and the number of trucks was 2,319,825, i.e. approximately 122 times fewer than passenger cars. The total number of those vehicles in 2018 is presented in Figure 2.1. Countries are sorted by the number of passenger cars. However, the Eurostat data did not cover the reports from Greece and Iceland.

Among passenger cars, the vehicles registered in Germany (17%), Italy (14%), France (11%) and the United Kingdom (11%) constitute the largest share. Poland is at the sixth place with the number of vehicles around 23.8 million. The smallest number of vehicles is registered in Malta (256,000), Kosovo (300,000) and Liechtenstein (29,000), which means less than 1% share in the total European automotive passenger car list. The situation is slightly different for heavy goods vehicles (HGV). Poland has the largest share (18%), followed by Turkey (10%), Spain (10%) and Germany (9%).

The total number of road vehicles strictly depends on the size of the country, in particular on the number of citizens; therefore, it is not a reliable indicator for environmental impact of road transport. A distribution of vehicles in smaller regions enable obtaining a more detailed insight into situation. The number of vehicles in the main administrative districts of each country is shown in Figure 2.2 for passenger cars and in Figure 2.3 for HGV (lorries and road tractors). General proportions are similar for these two categories of vehicles. For passenger cars, these are regions mainly located in Spain, France, Italy, Germany and Poland, and one region each in Great Britain and Turkey. In the case of HGV, a greater number can be seen in western Turkey.

The number of passenger cars per 1,000 inhabitants among European countries in 2018 is presented in Figure 2.4. The highest rate of 788 was in Liechtenstein. It is followed by the countries with rates in the range of 600–700, such us Luxembourg (676), Italy (646), Cyprus (629), Finland (629), Poland (617) and Malta (608). Interestingly, in the countries with the largest number of cars, the above-mentioned rate is not ascertained as highest, e.g. in Germany (567), France (478), the UK (743) and Spain (513). The lowest number of cars in 2018 was reported in Turkey (151), North Macedonia (200) and Romania (332). The data from Iceland and Kosovo were not published. The

DOI: 10.1201/9781003206149-2

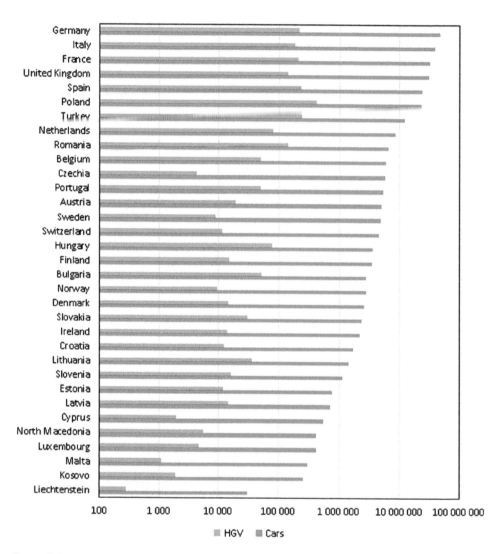

Figure 2.1 Total number of passenger cars and heavy goods vehicles (HGV) in 2018 (Eurostat 2020).

average factor is 498 cars per 1,000 of inhabitants. According to the data available in Eurostat, from 2010 to 2018, the rate increased from 448 to 498, respectively, which means an average annual increase of about 6 cars per 1,000 inhabitants. If the linear incremental trend was maintained (rate = 5.1268 year −9,858; $R^2 = 0.8621$; year in format 20##), then in 2030 the above-mentioned rate would be about 550 vehicles per 1000 inhabitants. There is a noticeable correlation between the number of vehicles per 1,000 inhabitants and the gross domestic product (GDP) per capita. The relationship is relatively best rendered ($R^2 = 0.459$) by the logarithmic trend line with the equation rate = 112.8 ln (GDP) −8.1905, where GDP is in million Euro per capita. Considering

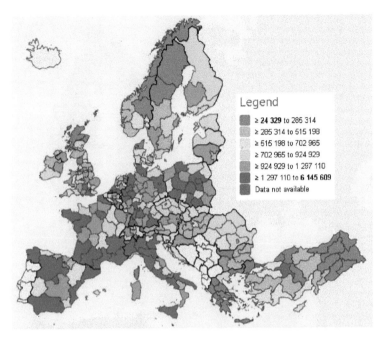

Figure 2.2 Number of passenger cars in national administrative main territories (Eurostat 2020).

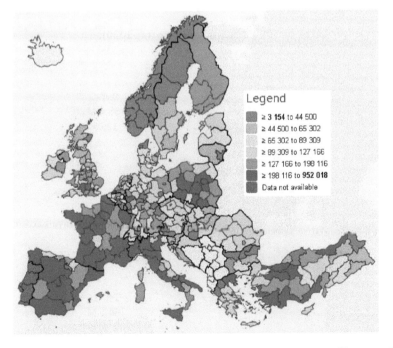

Figure 2.3 Number of HGV in national administrative main territories (Eurostat 2020).

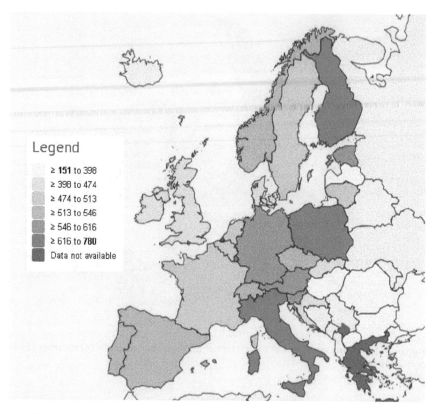

Figure 2.4 Passenger cars per 1000 inhabitants (Eurostat 2020).

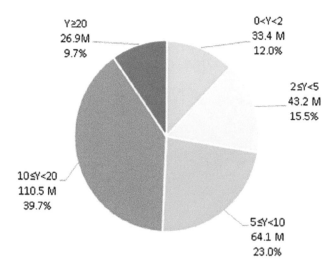

Figure 2.5 Number of cars in particular age category in Europe in 2018 (Eurostat 2020).

the statistically weak relationship above, it can be concluded that other factors also affect the number of vehicles. Taking into account the constant tightening of emission standards, shifting cars' production to more environment-friendly vehicles and socio-economic problems (for example, the global impact of COVID-19), a further increase in the number of passenger cars is difficult to predict.

The age of vehicles is an important factor influencing pollutant emissions. Figure 2.5 shows the percentage share in the EU of the individual age classes of passenger cars within reported in Eurostat vehicles. In the database, vehicles are grouped into five main categories: younger than 2 years (0–2), 2–5, 5–10, 10–20 and older than 20 years. On average, every eighth car was new (0–2). In general, the cars younger than 5 years (0–2 and 2–5) accounted for around a quarter of the total number of vehicles, which corresponds to the Euro 5–6 class regarding EES (European Emission Standard). The most numerous group of cars, approximately 110.5 million, belongs to age range 10–20 year, which corresponds to the Euro 2–4 class. In total, the cars older than 10 years account for practically half of the EU market, amounting to 137.4 million cars (Eurostat 2020).

The percentage share of individual age groups of vehicles, calculated individually for the majority of European countries, is shown in Figure 2.6. The highest percentage of new cars (up to 2 years) is in Ireland (29%) Luxembourg (24%), Denmark (23%) and Belgium (23%); in turn, the lowest percentage is in North Macedonia, Kosovo and Lithuania (2% each). However, there are still many old cars in use that do not meet the current stringent Euro emission standards. Cars older than 20 years (in 2018), meeting the maximum Euro 2 standard, have a large share in Finland (25%), Turkey (26%) and Estonia (29%). Poland is the undisputed leader in the percentage of old cars, where more than 35% of such cars are in circulation. A significant share of cars which are 10–20 years old is found in North Macedonia (68%), Lithuania (63%) and Czechia (61%).

The percentage share of different type of engines in Europe in 2018 year is shown in Figure 2.7. The category of hybrid vehicles (HYB) includes: hybrid electric-petrol and plug-in hybrid petrol-electric (ELC_PET_HYB_PI), hybrid electric-petrol (ELC_PET_HYB), plug-in hybrid petrol-electric (ELC_PET_PI), hybrid electric-diesel and plug-in hybrid diesel-electric (ELC_DIE_HYB_PI), hybrid diesel-electric (ELC_DIE_HYB) and plug-in hybrid diesel-electric (ELC_DIE_PI). The category of gas-powered vehicles (GAS) includes: liquefied petroleum gases (LPG), natural gas (NG), compressed natural gas (CNG) and liquefied natural gas (LNG). The category of biofuel vehicles (BIO) comprises: Bioethanol (BIOETH), Biodiesel (BIODIE) and Other (OTH). Further markings are as follows: PET – Petroleum products DIE – Diesel, ELC – Electricity, ALT – Alternative Energy and HYD_FCELL – Hydrogen and fuel cells.

The Euro emission standards have separate criteria for gasoline and combustion engines; however, for the currently applicable Euro 6d standard, there are no significant differences for most of the indicators. According to the data collected in 2018, the majority of vehicles in the EU were powered by gasoline (48.1%) and diesel (41.1%). The average engine displacement of gasoline units (<1,400 cm^3) was lower than in the case of diesel units (1,400 ÷ 1,900 cm^3). In many countries, gas-powered vehicles are much less popular. Fuel cell vehicles (822 items), electric vehicles (598,000 items) and biofuel (35,000 items) have a practically insignificant share.

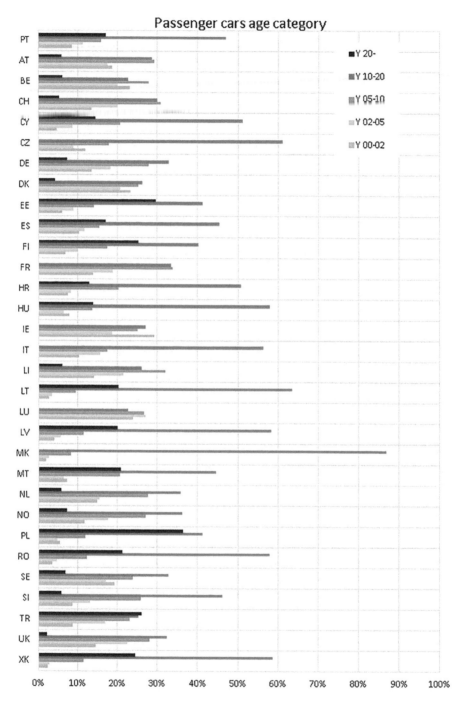

Figure 2.6 Percentage distribution of the age category of passenger cars in individual European countries (Eurostat 2020).

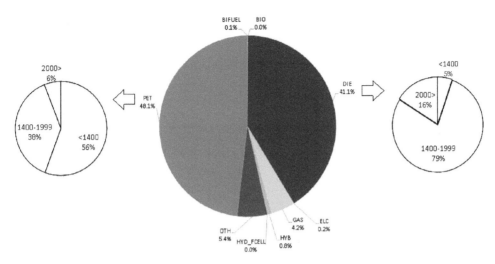

Figure 2.7 Percentage share of different type of engines in Europe in the year 2018 (central diagram) and engine displacement in cm^3 (diagrams on sides) for petroleum (left diagram) and diesel (right diagram) units (Eurostat 2020).

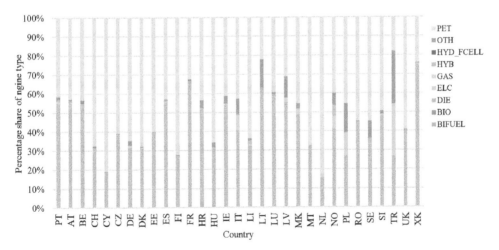

Figure 2.8 Percentage share of different types of engines in particular countries in 2018 (Eurostat 2020).

The situation in particular countries is shown in Figure 2.8. The highest percentage of petroleum vehicles in 2018 was in Cyprus (80%) and the Netherlands (82%) and the lowest in Turkey (18%), Lithuania (22%) and Kosovo (24%). Gas-powered engines can be found mainly in Turkey (27%), Poland (12%), Italy (7%) and Norway, while in other countries they are less popular.

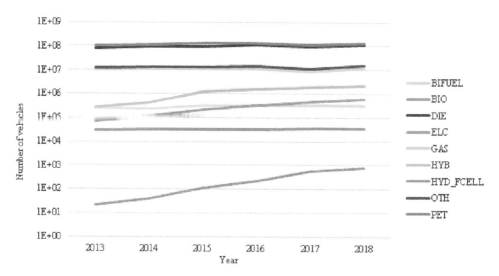

Figure 2.9 Changes in the number of vehicles in EU (Eurostat 2020).

Numerous scientific studies have been carried out to determine the environmental impact of the most popular types of engines, such as diesel, gasoline and LPG. In most cases, cars with diesel engines contribute the highest pollutant emissions.

In the studies conducted by Park et al. (2019), the high-market-share vehicles were selected for tests, according to the data provided by the Transportation Pollution Research Center in the National Institute of Environmental Research (NIER). All vehicles underwent transient chassis-dynamometer tests. The studies show that CO and NH_3 were dominant non-CO_2 pollutants emitted from gasoline and LPG cars, while NO_x, CO and non-methane hydrocarbons (NMHC) pollutants emissions dominated from diesel vehicles. The emissions of SO_2 from diesel vehicles were even up to 28 times higher than in the case of gasoline and LPG vehicles (Park et al., 2019).

LPG or CNG engines can be a good solution to reduce pollutant emissions. According to Johnson (2003), LPG should play a greater role in road transport because it is safer than most of alternative road transport fuels as well as is superior to most road transport fuels with respect to public health and environmental impact.

Hybrid cars have become increasingly popular in recent years; the percentage share of 0.75% is still insignificant, but the number of vehicles of this type is characterized by a visible increase in Figure 2.9. Currently, hybrid and electric cars have the greatest development potential. The average increase in period 2013–2018 of electric and hybrid cars and hydrogen fuel cell is equal, respectively, at 56% and 60%. Although an increase of hydrogen fuel cell cars is unquestionably high at 114%, their usage still corresponds to specialist applications (0.0001% share in total amount of vehicle).

The quantity of e-roads and motorways and average density of these roads (calculated together) per $1,000 km^2$ are presented in Figure 2.10. The weighted average value (regarding area) of road net density in the reported countries is equal to

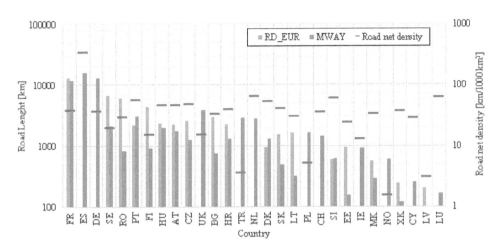

Figure 2.10 Road length and road net density for different countries in Europe (Eurostat 2020).

$25 \text{ km}/1{,}000 \text{ km}^2$. The lowest rate is in Norway due to the large non-urbanized area of the country. The highest density is in Estonia due to significant amount of motorway $(15{,}000.5 \text{ km})$ and relatively small country area $(45{,}000 \text{ km}^2)$.

2.2 EUROPEAN EMISSION STANDARD (EES)

The condition of the transport infrastructure may significantly contribute to air pollution, especially in urbanized areas. In order to reduce emissions, European countries have introduced the EES. The limits for selected pollution indicators have been introduced for individual vehicle types, such as: carbon monoxide (CO), hydrocarbons (THC), NMHC, oxides of nitrogen (NO_x), combined mass of total hydrocarbons and oxides of nitrogen (THC + NO_x), particulate matter (PM) and number of particles (PN). The list of individual indicators is presented in Table 2.1. In the case of diesel engines, the greatest change since Euro 1 occurred in PM. The value of the Euro 6d standard from 2020 in relation to Euro 1 from 1992 is 3%, and for CO and HC + NO_x it is 18%.

The graph in Figure 2.11 clearly shows how significant was the reduction of the two indicators until the introduction of Euro 5 standard for diesel engines. The change in the values of the indicators from the subsequent Euro 6- Euro 5 period compared to the change between Euro 1- Euro 5 is 0.37% and 8.11% for PM and HC + NO_x, respectively.

For the currently applicable Euro 6d standard, there are no significant differences in most indicators for diesel and gasoline engines, as presented in Figure 2.12. In the case of the CO standard for gasoline engines, it is 100% higher than diesel engines, whereas the NO_x index is 33% lower comparing these two types of engines. For the diesel engines, there is a specific standard for THC and VOC, while for gasoline engines, there is no defined standard of HC + NO_x.

Table 2.1 European Emission Standard for diesel and gasoline engines (Directive: 91/441/EEC, 2002/51/EC, 98/69/EC, EC 715/2007 2007, EU 459/2012 2012, EU 2016/646)

Engine	EES	Date of type approval	Date of first registration	PM g/km	PN #/km	CO g/km	THC g/km	VOC g/km	NOx g/km	HC+NOx g/km
Diesel	Euro 1	1992	1993	0.14	–	2.72	–	–	–	0.97
	Euro 2	1996	1997	0.08	–	1	–	–	–	0.7
	Euro 3	2000	2001	0.05	–	0.66	–	–	0.5	0.56
	Euro 4	2005	2006	0.025	–	0.5	–	–	0.25	0.3
	Euro 5a	2009	2011	0.005	–	0.5	–	–	0.18	0.23
	Euro 5b	2011	2013	0.0045	6×10^{11}	0.5	–	–	0.18	0.23
	Euro 6b	2014	2015	0.0045	6×10^{11}	0.5	–	–	0.08	0.17
	Euro 6c	–	2018	0.0045	6×10^{11}	0.5	–	–	0.08	0.17
	Euro 6d-Temp	2017	2019	0.0045	6×10^{11}	0.5	–	–	0.08	0.17
	Euro 6d	2020	2021	0.0045	6×10^{11}	0.5	–	–	0.08	0.17
Gasoline	Euro 1	1992	1993	–	–	2.72	–	–	–	0.97
	Euro 2	1996	1997	–	–	2.2	–	–	–	0.5
	Euro 3	2000	2001	–	–	2.3	0.2	–	0.15	–
	Euro 4	2005	2006	–	–	–	0.1	–	0.08	–
	Euro 5a	2009	2011	0.005	–	–	0.1	0.068	0.06	–
	Euro 5b	2011	2013	0.0045	6×10^{11}	–	0.1	0.068	0.06	–
	Euro 6b	2014	2015	0.0045	6×10^{11}	–	0.1	0.068	0.06	–
	Euro 6c	–	2018	0.0045	6×10^{11}	–	0.1	0.068	0.06	–
	Euro 6d-Temp	2017	2019	0.0045	6×10^{11}	–	0.1	0.068	0.06	–
	Euro 6d	2020	2021	0.0045	6×10^{11}	–	0.1	0.068	0.06	–

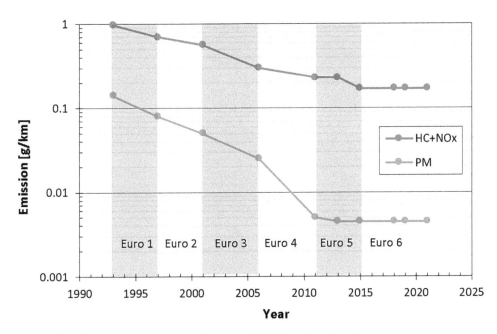

Figure 2.11 Changes od chosen coefficient of EES for diesel engines.

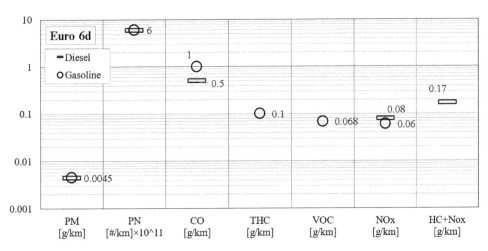

Figure 2.12 Differences in emission for diesel and gasoline engines according to the current Euro 6d standard.

Chapter 3

Detailed situation of road transport in Poland

Traffic jams are connected with the traffic in large cities. The annual international Traffic Index (TI) report describing the traffic situation in 416 cities from 57 countries was published by TomTom (2019). The list of the 19 most congested cities in the world as well as all Polish cities is presented in Table 3.1.

The basis of the ranking is the congestion level percentage, which describes what percent of additional time should be spent on commuting compared to normal

Table 3.1 Position of Polish cities in 2019 world traffic ranking (TomTom 2019)

No. in world rank	City	Country	Congestion Level (%)
1	Bengaluru	India	71
2	Manila	Philippines	71
3	Bogota	Colombia	68
4	Mumbai	India	65
5	Pune	India	59
6	Moscow region (oblast)	Russia	59
7	Lima	Peru	57
8	New Delhi	India	56
9	Istanbul	Turkey	55
10	Jakarta	Indonesia	53
11	Bangkok	Thailand	53
12	Kyiv	Ukraine	53
13	Mexico City	Mexico	52
14	Bucharest	Romania	52
15	Recife	Brazil	50
16	Saint Petersburg	Russia	49
17	Dublin	Ireland	48
18	Odessa	Ukraine	47
19	Łódź	Poland	47
22	Kraków	Poland	45
27	Poznań	Poland	44
37	Warsaw	Poland	40
41	Wrocław	Poland	39
71	Bydgoszcz	Poland	34
82	Gdańsk, Gdynia and Sopot	Poland	33
122	Szczecin	Poland	30
172	Lublin	Poland	27
191	Białystok	Poland	26
267	Bielsko-Biała	Poland	21
320	Katowice urban area	Poland	19

DOI: 10.1201/9781003206149-3

conditions with standard traffic. The worst situation is in Bengaluru (India) and Manila (Philippines), where drivers spend an average of 71% of the extra time in traffic jams. The next busiest cities are Bogota (Colombia) at 68%, Mumbai (India) at 65%, Pune (India) at 59% and Moscow (Russia) at 59%. Among the ten most congested cities in the world, four are from India.

Out of 416 cities on the list, 12 are Polish cities, including 7 in the top 100 of TI. Apart from the city of Bielsko-Biała, all of them are voivodeship cities. The remaining voivodeship cities that were not on the TI list are: Gorzów Wielkopolski, Opole, Rzeszów, Kielce and Olsztyn. For this reason, it is essential for the main Polish cities to analyze the dependence of the pollution level on various communication indicators, such as the total number of vehicles, the number of vehicles of individual types, the number of vehicles in age categories, and the condition of road infrastructure in these urban poviats.

3.1 DATA SOURCE

The data for the analysis of the vehicles condition and road infrastructure were taken from the local data bank of the Statistics Poland in the field of transport and communication (Statistics Poland 2020a):

- vehicles in total (2002–2019),
- vehicles by fuel (2015–2019),
- vehicles by age (2015–2019),
- public poviat roads – indicators (2005–2019).

Total vehicles has been included since 2002, as since then the data has come from the administrative data source (poviat databases on registered vehicles). Earlier data came from T03 reports (Statistics Poland 2020a) that were quantitatively different from the administrative data and were therefore omitted. The data for the whole of Poland were used for general analyzes of data from the country, and for individual voivodeship cities, the data from urban poviats were used.

The data on air quality were collected from the measurement data bank of the National Environmental Monitoring network of the Chief Inspectorate for Environmental Protection (GIOS 2020). The focus was on four air quality indicators: PM_{10} and $PM_{2.5}$ particulate matter, NO_x nitrogen compounds, and benzo(a)pyrene (measured with PM_{10}). The data from the following cities specified in Table 3.2 were used. The cities on the list of the most congested cities according to the TomTom ranking (TomTom 2019) are marked with a gray background.

3.2 TOTAL NUMBER OF VEHICLES

The total number of vehicles in Poland over two decades is shown in Figure 3.1. According to the latest data published by the Statistics Poland, covering the year 2019, there were 34,391,507 vehicles in the inventory, including 24,360,166 passenger cars (71% of the total) and 3,436,184 trucks (10%) (Statistics Poland 2020a). The fleet of trucks is a significant contributor to traffic pollution. Their number is much smaller,

Table 3.2 Analysed Polish city

City	Voivodeship	City code
Białystok	podlaskie	PdBial
Bydgoszcz	kujawsko-pomorskie	KpByd
Gdańsk	pomorskie	PmGda
Gorzów Wielk	lubuskie	LuGorz
Katowice	śląskie	SlKato
Kielce	świętokrzyskie	SkKiel
Kraków	małopolskie	MpKrak
Lublin	lubelskie	LbLub
Łódź	łódzkie	LdLodz
Olsztyn	warmińsko-mazurskie	WmOls
Opole	opolskie	OpOpole
Poznań	wielkopolskie	WpPozn
Rzeszów	podkarpackie	PkRzesz
Szczecin	zachodniopomorskie	ZpSzcz
Warszawa	mazowieckie	MzWar
Wrocław	dolnośląskie	DsWroc

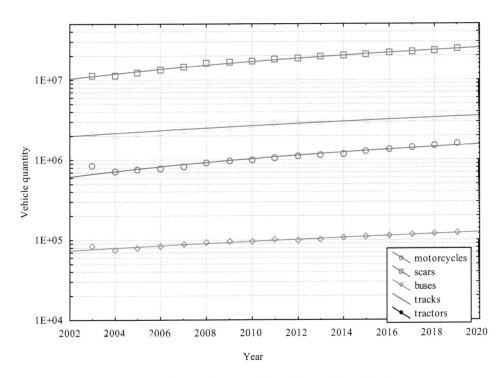

Figure 3.1 Overall quantity of general types of vehicles in Poland during years.

but the vehicles have a much greater engine power, which translates into higher fuel consumption. For a number of years, there has been a statistically significant linear upward trend, which is confirmed by the high coefficients of determination presented in Table 3.3. With the current intensity of the upward trend, there may be approximately

Table 3.3 Statistical Parameters of Data Series from Figure 3.1

Vehicle	Linear equation	R^2
Motocycles	$y = -1.041 \times 10^8 + 52{,}310.919\ x$	0.9544
Cars	$y = -1.642 \times 10^9 + 8.255 \times 10^5\ x$	0.9919
Buses	$y = -5.561 \times 10^6 + 2{,}814.713\ x$	0.9719
Trucks	$y = -1.719 \times 10^8 + 86{,}855.061\ x$	0.9767

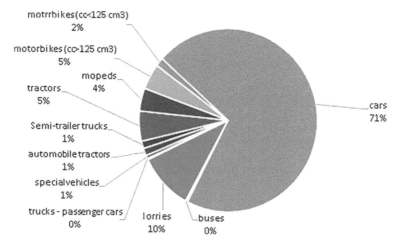

Figure 3.2 Vehicle types in Poland in 2019.

33.5 million passenger cars in 2030 and 4.4 million trucks in the year 2030. The average annual increase in passenger cars is almost 900,000, while the increase in trucks is about ten times lower. The percentage share of other types of vehicles is significantly lower in total. All motorcycle categories have an 11% market share and agricultural tractors account for 5% (Figure 3.2).

3.3 AGE STRUCTURE OF VEHICLES

The vehicles aged 16–20 (20%) had the largest share of passenger cars on the market in 2019. The next largest group were cars aged 12–15 (16%) and 21–25 (14%). New cars, up to the first, second and third years of the year constituted 4%, 2% and 2%, respectively. Cars older than 31 years (manufactured before 1989) do not meet the emission standards, but in many cases they are vintage and unusual cars, or used for traveling shorter distances. Their impact on the environment is difficult to determine. It is not possible to keep records of the distance traveled or the operating time of the combustion engine. It is estimated that the environmental impact of cars older than 31 years old should be lower than that of the cars, e.g. 12–15 years old, which have the same percentage share. The passenger cars older than 12 years make up approximately 3/4 of the current market (2019).

The age structure of trucks is very similar to that of passenger cars. There is a slightly smaller share of vehicles in the categories of 12–15, 16–20 and 21–25 years, and higher for the age category ≥31 (Figure 3.3).

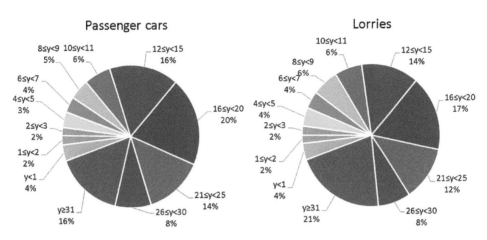

Figure 3.3 The percentage share of individual vehicle years in Poland in 2019 among passenger cars and trucks.

The changes in the number of vehicles over 5 years in individual age classes are presented in the graphs in Figure 3.4. The change in the age structure of passenger cars is not proportional for all age categories. Among passenger cars, there is a noticeable increase in older cars compared to new cars. For example, between 2015 and 2019, 301,709 new cars were registered (<1 year). Due to traffic pollution, it would be beneficial if older and sub-standard cars were systematically withdrawn from the market. In 2019, the cars older than 27 years old (approved until 1992, registered until 1993) can be considered such passenger cars. In the age classes 26–30 and ≥31, there is a continuing upward trend in the number of these vehicles. Lorries and other types of vehicles are subject to different emission standards, but here too, the increase in the number of cars from older vehicles is clearly visible.

It is difficult to compare the size of the different age groups because of the different age ranges. The largest increase in vehicles was recorded in the group of older vehicles, generally over 12 years of age (e.g. 21–25: 733,705 units). For the 6–11 age classes, there was a surprising decrease in the number of vehicles. On the other hand, the relative change in the number of vehicles in age groups may be decisive (Table 3.4). The results indicate that the groups up to 3 years of age are characterized by a significant relative increase.

3.4 DIVISION OF VEHICLES BY TYPE OF FUEL

The percentage share of passenger cars by type of fuel used practically remained unchanged from 2015 to 2019 (Figure 3.5). There was a slight decrease in gasoline-powered cars (−3%) and an increase in diesel cars (2%) as well as other vehicles (1%). The percentage of LPG-fueled vehicles remains unchanged. From the data available in the Statistics Poland, it is not possible to analyze which age groups of cars were fueled by a given type of fuel and, thus, assign the appropriate EES. Only an average value of 54% for petrol and 31% for diesel can be taken (Statistics Poland 2020a).

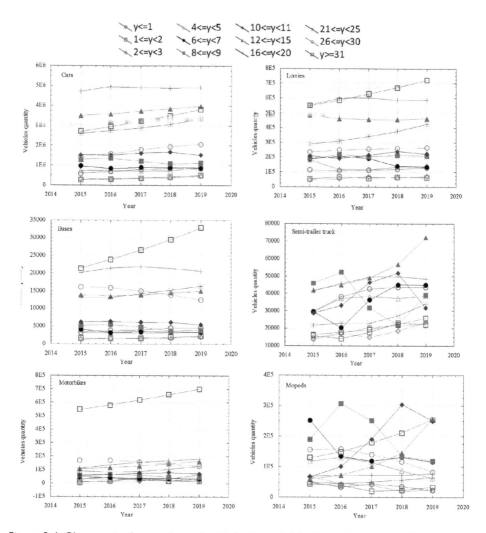

Figure 3.4 Changes in the number of vehicles in individual age classes over 5 years.

In Figure 3.6, the number of vehicles by fuel type for the considered cities is provided. In all cities, since 2016, there has been a noticeable increase in the number of cars with fuel referred to as "other". It is related to the popularization of electric cars.

Figure 3.7 shows the percentage share in 2019 of passenger cars produced during the period of application of individual emission standards for gasoline and diesel engines taken together. For age categories covering several years, the middle year was adopted as the date representing the given age category. The largest percentage of vehicles was sub-standard. More than 75% of vehicles can be classified below the Euro 5a standard (2009 date of type approval, 2011 date of first registration) and have emissions of $PM \geq 0.005$ g/km, $CO \geq 0.5$ g/km, $NO_x \geq 0.18$ g/km and $HC + NO_x \geq 0.23$ g/km. This is an important issue because, until the introduction of the Euro 5 standard, selected permitted indicators were an order of magnitude higher.

Table 3.4 Comparison of the relative changes in the number of vehicles in individual age categories of passenger cars

Age category	2015	2019	Change	% change (%)
$y<1$	600,589	902,298	301,709	50
$1 \leq y<2$	284,884	510,321	225,437	79
$2 \leq y<3$	318,535	462,163	143,628	45
$4 \leq y<5$	729,141	861,738	132,597	18
$6 \leq y<7$	959,455	864,135	−95,320	−10
$8 \leq y<9$	1,302,196	1,120,892	−181,304	−14
$10 \leq y<11$	1,526,177	1,510,642	−15,535	−1
$12 \leq y<15$	3,515,910	3,983,987	468,077	13
$16 \leq y<20$	4,710,323	4,914,323	204,000	4
$21 \leq y<25$	2,605,976	3,339,681	733,705	28
$26 \leq y<30$	1,458,589	2,074,164	615,575	42
$y \geq 31$	2,711,648	3,815,822	1,104,174	41

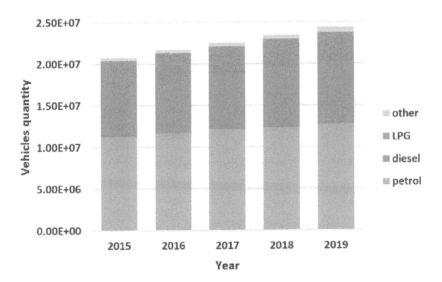

Figure 3.5 Percentage of passenger cars by type of fuel used.

3.5 ASSESSMENT OF LONG-TERM CHANGES IN ROAD INDICATORS TO THE AVERAGE LEVEL OF POLLUTION

Theoretically, the concentration of pollutants should be proportional to the number of vehicles. In reality, however, the above-mentioned hypothesis is not confirmed, as in the last 10 years there has been a significant tightening of the emission standards for new cars. Figure 3.8 shows the relationship between the concentration of PM_{10} and the number of passenger cars for individual voivodeship cities in 2009–2018. The results of most cases are not significant for the adopted significance level of 0.05, except for Gorzów Wielkopolski, Kraków and Wrocław. In these three cases, the opposite tendency can be seen: the decrease in PM_{10} concentration with the increase in the number of vehicles is seen. Trend lines have quite a high coefficient of determination R^2.

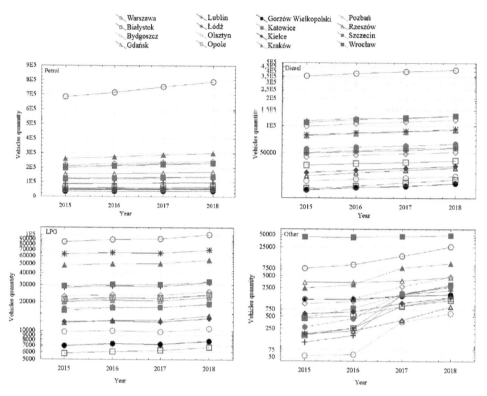

Figure 3.6 Number of vehicles in particular fuel types in selected cities.

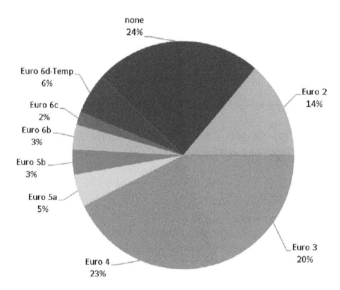

Figure 3.7 Percentage share in 2019 of passenger cars produced during the period of application of individual emission standards.

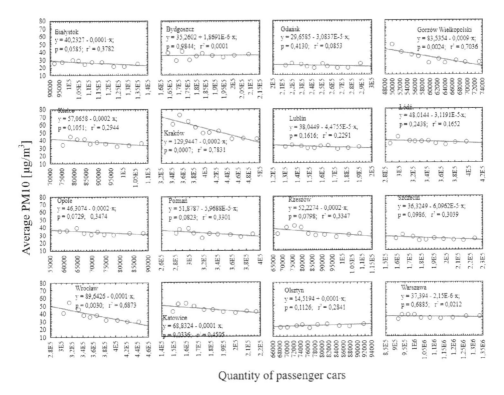

Quantity of passenger cars

Figure 3.8 Relationship between car quantity and PM₁₀ concentration in 2009–2018 for particular cities.

The conducted analyses also show no correlation between the number of old cars (over 10 years old) and the concentration of PM₁₀ particulate matter. Cars older than 10 years do not meet the Euro 5 standard, from which there has been a significant tightening of emission standards (Figure 3.9). The statistical database of the Statistics Poland covers only 4 full years of published data on the age category of cars (Statistics Poland 2020a).

A greater number of hardened public urban roads facilitates efficient movement; however, the statistics presented below do not confirm a significant relationship between the density of roads and the concentration of PM₁₀ (Figure 3.10).

3.6 AIR QUALITY IN 2019 TAKING INTO ACCOUNT THE TYPE OF MEASURING STATION

Communication contributes significantly to air pollution. For this reason, this chapter presents in a synthetic way to take detailed measurement data for the type of SEM stations for 2019. The following parameters were taken into account: PM₁₀ (µg/m³), PM₂.₅ (µg/m³), SO₂ (µg/m³), NOₓ (µg/m³) and CO (mg/m³).

There are three types of measuring stations, referred to as background, communication and industrial. In addition, these stations are located in the urban, suburban

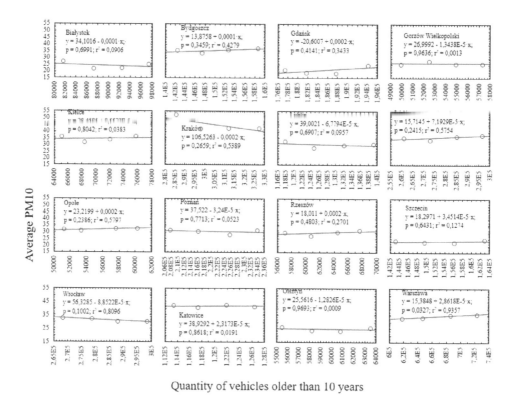

Figure 3.9 Relationship between the number of old cars (>10 years) and PM₁₀ concentration in 2015–2018 for particular cities.

and extra-urban zones. In the case of individual air pollutants, the number of stations broken down by station type and area is presented in Table 3.5. At the end of 2019, a total of 262 measuring stations were active, including 237 background (189 urban, 24 suburban, non-urban 24), 16 communication and 9 industrial. Detailed data of communication stations are given in Table 3.6.

For the analysis, the hourly data from the archive of the Chief Inspectorate for Environmental Protection containing measurement results from 2019 for all active stations in Poland were collected. Therefore, these are average and representative values for all cities in the country and should not be directly related to a specific location. The data has been limited to even hours.

In most cases, the concentration level of the considered indicators is higher for urban communication stations than for the urban background. Only for SO_2 and benzo(a) pyrene (BaP), a higher concentration was observed for the urban background. These pollutants more closely reflect the sources from the domestic and utilities sector, such as real estate heating. Solid fuel boilers and stoves are still installed in many places in Poland.

It is also evident in the charts that the concentration of individual indicators is lower in the summer. In Poland, the heating season lasts on average from mid-October to

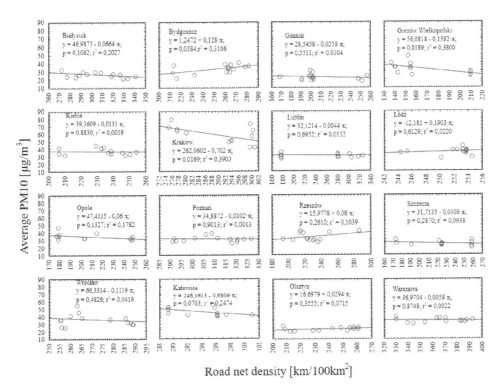

Figure 3.10 Relationship between the number of old cars (>10 years) and PM$_{10}$ concentration in 2015–2018 for particular cities.

Table 3.5 Number of particular measurement stations for pollutants according to data for 2019

Station type	Area type	PM$_{10}$	PM$_{2.5}$	SO$_2$	NO$_x$	CO
Background	Urban	108	44	86	95	52
	Suburban	10	7	7	13	3
	Extra-urban	4	2	18	20	2
Communicational	Urban	12	9	8	16	14
Industrial	Urban	4	1	4	4	2
	Suburban	-	-	1	-	-
	Extra-urban	1	-	1	1	-

mid-April. Pursuant to the ordinance of the Minister of Economy of 2007, the heating season is a period in which weather conditions require continuous heat supply to heat the facilities. The legal provisions have no specified date or top-down criteria for determining the boundary periods. It depends on the individual preferences of the residents, the current meteorological conditions and the technical condition of the property.

Table 3.6 Addresses of communication stations active in 2019

Voivodeship	City	Station code	Latitude φ N (WGS84)	longitude λ E (WGS84)
Dolnośląskie	Wrocław	DsWrocAlWisn	51.086225	17.012689
Kujawsko-Pomorskie	Bydgoszcz	KpBydPlPozna	53.121764	17.987906
	Grudziądz	KpGrudPilsud	53.493550	18.762139
	Toruń	KpToruKaszow	53.017628	18.612808
	Włocławek	KpWloclOkrze	52.658467	19.059314
Łódzkie	Łódź	LdLodzJanPaw	51.754613	19.434925
Małopolskie	Kraków	MpKrakAlKras	50.057678	19.926189
		MpKrakDietla	50.057447	19.946008
	Tarnów	MpTarRoSitko	50.018253	20.992578
Mazowieckie	Warszawa	MzWarAlNiepo	52.219298	21.004724
Podkarpackie	Rzeszów	PkRzeszPilsu	50.040675	22.004656
Śląskie	Bielsko-Biała	SlBielPartyz	49.802075	19.048610
	Częstochowa	SlCzestoArmK	50.817217	19.118997
	Katowice	SlKatoPlebA4	50.246795	19.019469
Zachodniopomorskie	Koszalin	ZpKoszArKraj	54.193986	16.172544
	Szczecin	ZpSzczPils02	53.432169	14.553900

3.6.1 PM$_{10}$

The permissible concentration of particulate matter PM$_{10}$ according to the Regulation of the Minister of the Environment on the levels of certain substances in the air (*Journal of Laws* 2019 pos. 1931, *Journal of Laws* 2012 pos. 1031) is 50 µg/m^3 for the daily average. The frequency histogram of the individual concentrations is shown in Figure 3.11. The most common concentrations for the background, communication and industrial stations are 12, 16 and 12 µg/m^3; 90% of all hourly results are below 50 µg/m^3.

Figure 3.12 shows the results broken down into hours and months. There is a noticeable reduction in the value of pollutants in the summer months. In the case of an hourly statement, there is no specific time for which the predominant concentration would be (Table 3.7).

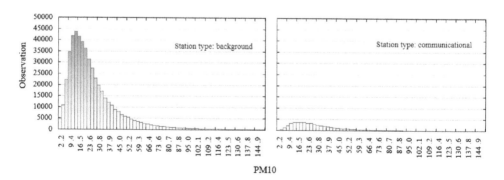

Figure 3.11 Histogram of PM$_{10}$ depending on the type of station.

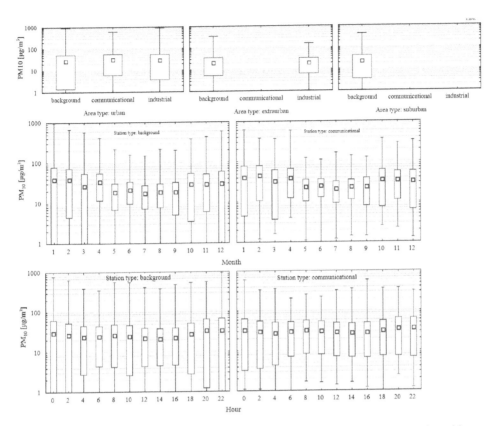

Figure 3.12 Concentration of PM$_{10}$ depending on station and area type, month and hour.

Table 3.7 Basic statistics for concentration of PM$_{10}$ depending on station and area type

Station type	Area type	Mean	Median	Minimum	Maximum	Percentile (10)	Percentile (90)	Std. Dev.
Background	Urban	27.0	20.3	0.0	951.1	8.1	51.5	25.6
Industrial	Extra-urban	19.4	16.3	0.0	158.2	6.6	36.0	13.3
Communicational	Urban	31.0	24.5	0.6	613.9	10.3	57.8	24.9
Background	Suburban	22.8	17.7	0.0	415.8	7.2	43.4	19.2
Background	Extra-urban	20.2	16.9	0,0	349.0	6.3	38.0	14.8
Industrial	Urban	28.9	22.0	0.4	920.1	9.0	56.0	25.2

3.6.2 PM$_{2.5}$

The permissible concentration of particulate matter PM$_{2.5}$ according to the Regulation of the Minister of Environment on the levels of certain substances in the air (*Journal of Laws* 2019 pos. 1931, *Journal of Laws* 2012 pos. 1031) is 20 µg/m^3 for annual averaging.

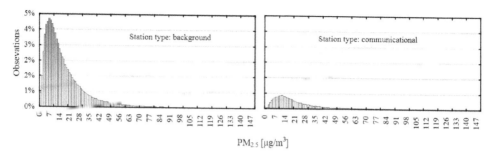

Figure 3.13 Histogram of PM$_{2.5}$ depending on the type of station.

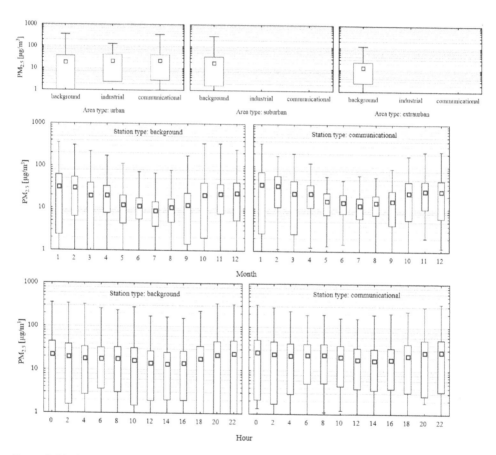

Figure 3.14 Concentration of PM$_{2.5}$ depending on station and area type, month and hour.

In 2019, this condition was met, as the average annual concentration was 18.4 μg/m^3 for the urban background. The frequency histogram of the individual concentrations is shown in Figure 3.13. The most common concentrations for the background, communication and industrial stations are 7, 11 and 3 μg/m^3, respectively (Figure 3.14, Table 3.8).

Table 3.8 Basic statistics for concentration of PM$_{2.5}$ depending on station and area type

Station type	Area type	Mean	Median	Minimum	Maximum	Percentile (10)	Percentile (90)	Std. Dev.
Background	Urban	18.7	13.6	0.0	351.8	5.0	37.4	17.9
Communicational	Urban	22.7	17.2	0.0	346.0	7.1	44.1	19.9
Industrial	Urban	21.7	15.4	3.0	127.7	5.0	46.0	19.2
Background	Suburban	18.4	13.6	0.2	293.8	4.9	36.8	16.7
Background	Extra-urban	12.6	10.5	0.0	122.4	2.6	24.8	10.2

3.6.3 SO$_2$

The permissible concentration of particulate matter SO$_2$ according to the Regulation of the Minister of Environment on the levels of certain substances in the air (*Journal of Laws* 2019 pos. 1931, *Journal of Laws* 2012 pos. 1031) is 350 μg/m^3 for hourly averaging, 125 μg/m^3 for daily averaging and 20 μg/m^3 for annual averaging. In 2019, this condition was met because the average annual concentration was 5.2 μg/m^3 for urban background and 4.1 for communication stations. The frequency histogram of the individual concentrations is shown in Figure 3.15. The most common concentration for the background and communication station ranges from 0.5 to 1 and μg/m^3.

When analyzing the average results for individual months, in the case of SO$_2$, there is also a noticeably higher concentration of pollutants in the winter. The average concentration in individual hours is relatively stable throughout the day. The maximum concentrations occurred in the urban area for the background 453.6 μg/m^3 and the communication station 439.1 μg/m^3. In the remaining areas (excluding the suburban industrial zone), the maximum recorded concentrations were over two times lower (Figure 3.16, Table 3.9).

3.6.4 NO$_x$

Nitrogen oxides can be considered one of the most characteristic pollutants for transport. The permissible concentration of particulate matter NO$_x$ according to the Regulation of the Minister of Environment on the levels of certain substances in the air (*Journal of Laws* 2019 pos. 1931, *Journal of Laws* 2012 pos. 1031) is 30 μg/m^3 for annual

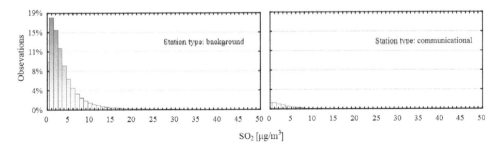

Figure 3.15 Histogram of SO$_2$ depending on the type of station.

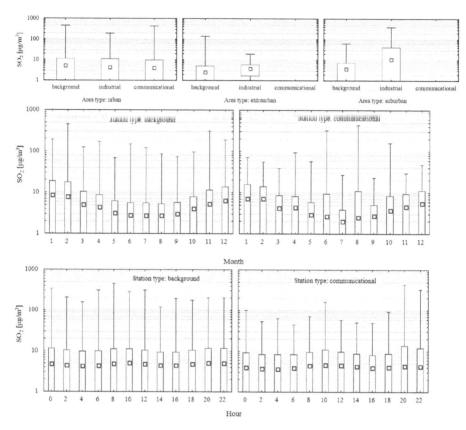

Figure 3.16 Concentration of SO₂ depending on station and area type, month and hour.

Table 3.9 Basic statistics for concentration of SO₂ depending on station and area type

Station type	Area type	Mean	Median	Minimum	Maximum	Percentile (10)	Percentile (90)	Std. Dev.
Background	Urban	5.2	3.5	-0.5	453.6	1.1	10.5	6.5
Industrial	Extra-urban	3.8	3.4	0.0	19.3	1.8	6.4	2.1
Background	Extra-urban	2.5	1.9	0.0	139.5	0.6	5.0	2.6
Communicational	Urban	4.1	2.8	0.0	439.1	0.6	8.6	5.7
Background	Suburban	3.8	2.7	0.0	63.2	0.7	7.9	3.7
Industrial	Urban	4.4	2.9	−0.1	193.0	0.9	8.7	6.4
Industrial	Suburban	10.7	3.4	0.0	385.5	1.5	15.0	31.0

averaging. In 2019, this condition was met for the urban background and the industrial area, as the average annual concentration was 20.8 and 24.8 µg/m³, respectively. In the case of communication stations, the situation is much worse, where the average annual concentration was as high as 80.8 µg/m³. Although the occasional maximum concentration for the communication station and the urban background are similar (around 1,500), the 90 percentile in the vicinity of communication routes is signifi-cantly higher.

In the case of the hourly graph, higher NO_x levels occur at 8 a.m. and 8 p.m. The first increase is likely due to the morning traffic peak. Seasonal sources of pollution also affect pollution levels. For all zones, the concentration is lower during the summer (Figures 3.17 and 3.18, Table 3.10).

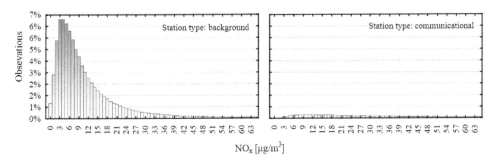

Figure 3.17 Histogram of NO_x depending on the type of station.

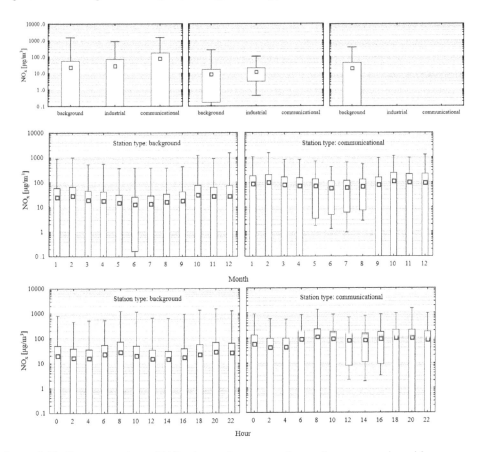

Figure 3.18 Concentration of NO_x depending on station and area, month and hour.

Table 3.10 Basic statistics for concentration of NO$_x$ depending on station and area type

Station type	Area type	Mean	Median	Minimum	Maximum	Percentile (10)	Percentile (90)	Std. Dev.
Background	Urban	23.5	14.7	0.0	1532.2	5.5	46.2	32.8
Industrial	Extra-urban	11.9	9.6	0.4	106.4	3.9	?? ?	8.7
Background	Extra-urban	8.5	6.2	0.0	266.0	2.1	16.9	8.3
Communicational	Urban	80.8	50.9	0.9	1530.2	13.5	184.1	89.1
Background	Suburban	19.7	13.1	0.0	381.7	5.0	39.5	22.0
Industrial	Urban	28.1	16.2	0.0	863.0	5.6	56.5	43.6

3.6.5 CO

The permissible concentration of particulate matter (CO) according to the Regulation of the Minister of Environment on the levels of certain substances in the air (*Journal of Laws* 2019 pos. 1931, *Journal of Laws* 2012 pos. 1031) is 10 mg/m^3 for an 8-hour averaging. In 2019, this condition was met for all types of areas. Even the maximum values are lower than the permissible limit and amounted to 5.5, 6.2 and 3.5 mg/m^3 for urban background, communication and industrial stations, respectively (Figures 3.19 and 3.20, Table 3.11).

3.7 SUMMARY

Communication contributes significantly to air pollution. However, the analyzed data from the whole country shows that municipal communication stations are characterized by a higher concentration of pollutants compared to background stations with: PM$_{10}$ 17%, PM$_{2.5}$ 24%, CO 36% and NO$_x$, by as much as 288%. However, in the case of SO$_2$, the concentration for communication stations is lower than the background stations by 12%. Sulfur dioxide is a pollutant typical of the coal combustion process carried out by entities in the municipal sector.

The proportion of the share of individual age classes has also changed over the years. There was a significant percentage increase in the number of new cars, but old cars that do not meet any EURO emission standards are still not being deregistered or withdrawn from use. In Poland, in 2019, about 3/4 of the market for passenger cars

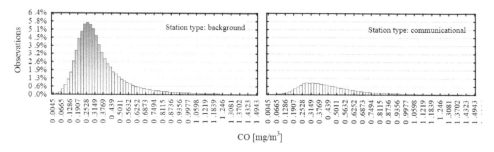

Figure 3.19 Histogram of CO depending on the type of station type, month and hour.

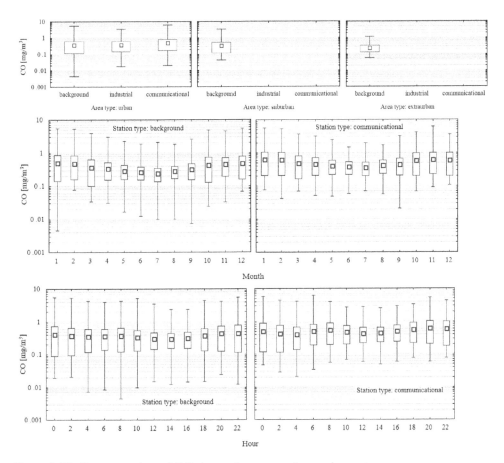

Figure 3.20 Concentration of CO depending on station and area type.

Table 3.11 Basic statistics for concentration of CO depending on station and area type

Station type	Area type	Mean	Median	Minimum	Maximum	Percentile (10)	Percentile (90)	Std. Dev
Background	Urban	0.372	0.309	0.004	5.526	0.181	0.609	0.256
Communicational	Urban	0.496	0.415	0.021	6.212	0.231	0.833	0.322
Background	Suburban	0.334	0.283	0.044	3.470	0.163	0.550	0.212
Background	Extra-urban	0.232	0.214	0.059	1.153	0.119	0.365	0.102
Industrial	Urban	0.387	0.325	0.018	3.552	0.190	0.647	0.239

comprised the vehicles older than 12 years. On the other hand, the increase in the number of vehicles powered by fuels other than oil, gasoline and gas since 2016 is a good sign. However, their share of the total number of vehicles on the market is small, around 3%.

Chapter 4

Traffic-related pollution in Lublin

4.1 TRAFFIC-RELATED POLLUTION ISSUES

Air toxics constitute a huge and diversified group of pollutants known or suspected to induce harmful effects in people, given sufficient exposure. These include cancer, the impact on the tissue and organ development, and detrimental effect on the reproductive neurologic, immune and respiratory systems. Air toxics are generated by various outdoor and indoor sources, threatening large groups of people. Although the ambient levels of these compounds are usually low, they may cause public health concerns. Due to the low ambient levels, the techniques and tools used for the assessment of specific effects of air toxics on health are limited.

EPA also has to regulate or consider regulating the air toxics originating, at least partially, from motor vehicles (mobile source air toxics MSATs); hence, standards for fuels and vehicle emissions are set. Nevertheless, EPA took no specific regulatory action, since at that time the rules for sulfur reduction in gasoline and diesel aimed at particulate matter (PM) reduction were thought to simultaneously decrease several other MSATs (US EPA 2019). Afterwards, EPA identified 8 MSATs which – according to their reported toxicity and emissions – were most harmful to the human health, namely benzene, formaldehyde, 1,3-butadiene, naphthalene, acrolein, polycyclic organic matter, as well as diesel PM, and diesel exhaust organic gases. Additionally, EPA mandated reducing the content of benzene in gasoline and the content of hydrocarbons (MSATs included) in the exhaust gases (US EPA 2019).

The aerosol pollutants present in air may originate from natural or anthropogenic sources. The latter include the fuel combustion processes as well as transport and communication. When considering the sizes of emitted particles, it is commonly believed that the combustion processes in domestic furnaces are mainly a source of coarse particles (>2.5 μm), whereas internal combustion engines emit fine particles (<2.5 μm) (Ketzel et al. 2007, Karjalainen et al. 2014). The substances emitted from vehicle engines as well as the emissions from other sources and secondary emissions, e.g. caused by abrasion of tires, brake pads and road surface or resuspension from road surface, significantly contribute to an increased concentrations on roads and their vicinity. As a result, the quality of air inhaled by the road users is significantly deteriorated. Apart from the $PM_{2.5}$ and PM_{10} particles, road transport is also the source of such pollutants as nitrogen oxides, carbon oxide and volatile organic compounds.

According to the studies by Paasonen et al. (2013), the main sources of emission of particles with size below 300 nm include the road transport (over 60%), maritime and

DOI: 10.1201/9781003206149-4

air transport (19%), as well as boilers and furnaces (13%). Combustion engines, both with spark and compression ignition, are mainly the source of particles with sizes from 10 to 1,000 nm. It is estimated that diesel engines, in comparison with spark ignition engines, emit much greater amounts of fine particles. On the basis of the studies conducted by Maricq et al. (1999), the particle size distribution is log-normal with their greatest number in the range of 60–100 nm. Similar ranges of particle sizes emitted from combustion engines are also reported in more recent publications (Karjalainen et al. 2014, Goel and Kumar 2015).

In line with the GAINS (*Greenhouse gas – Air Pollution Interactions and Synergies*) model, the aerosol particles with sizes below 100 nm constitute approximately 84% of the total amount of particles in urban air (Paasonen et al. 2013). Such a high percentage of UFP in the total number of particles in the air was confirmed by numerous studies conducted in many cities of Europe, Asia and Australia (Dall'Osto et al. 2010, Kumar et al. 2008, Birmili et al. 2013, Mönkkönen et al. 2005). The studies conducted by Wu et al. (2008) in Beijing (one of the largest cities in the world, with over 2.5 million registered vehicles in 2005) pertaining to the particle with sizes from 3 nm to 10 μm indicated that their mean annual concentration amounted to 32,800 particles/cm^3. For the particles with size ranges of 3–20 nm, 20–100 nm and 0.1–1 μm, these concentrations amounted to 9,000 particles/cm^3 (27.4%), 15,900 particles/cm^3 (48.5%) and 7,800 particles/cm^3 (23.8%). It was shown that the concentrations of particles with the size of 20–100 nm were best correlated with the road traffic intensity. The share of UFP in total concentration of particles reached 75.9%. In the case of other cities of Europe and around the world, the share of UFP in the total concentration of particles is as follows: Alkmaar 70.9% (Ruuskanen et al. 2001), Erfurt 68.3% (Ruuskanen et al. 2001), Leipzig 90.1% (Wehner and Wiedensohler 2003), Pittsburg 90.0% (Stanier et al. 2004) and Atlanta 91.8% (Woo et al. 2001). In turn, the studies conducted in Brisbane, Australia, over the period of 11 years indicated that the share of UFP and nanoparticles (<50 nm) in the total number of particles measured in the air amounted to 82%–89% and 60%–70%, respectively (Mejía et al. 2008). According to other studies, the share of particles smaller than 300 nm in the total number of particles in the air close to communication routes may reach even 99% (Kumar et al. 2009). Generally, it can be stated that vehicular road traffic significantly determines the spatial distribution of UFP concentrations in urban agglomerations.

The studies conducted in an urban area of Dresden, Germany, confirmed that the concentrations of particles with sizes 5–300 nm are significantly correlated with the occurrence of urban sources of particle emissions and that the variability of road traffic volume is the cause of the highest fluctuations of these particles (Birmili et al. 2013). Especially large changes were observed for the concentrations of particles in the range of 5–20 nm. These studies also indicated that the photochemical processes play a major role in the nucleation and conversions of UFP. Larger particles in the range of 300–800 nm undergo relatively low changes in time and space, creating relatively homogeneous background.

High concentrations of aerosol particles are a factor that contributes to the low air quality in cities. The traffic-related emissions, aside from the emissions from the combustion of fuel for heating purposes, seriously reduce the quality of air in cities. This predominantly concerns the locations in vicinity of routes with heavy traffic. Epidemiological studies report the adverse effect of the traffic-related emissions on human health, especially of pedestrians and commuters (drivers and passengers). The exposure to the traffic-related pollution may have a negative effect on the consequences on the health of pedestrians. The study on this topic is of great importance in the cities of Central and Eastern Europe, taking into account the largely outdated transportation fleet and seasonal variability of weather conditions.

One of the essential parameters that determines the quality of air in cities includes the concentration of particles suspended in ambient air. These aerosol particles are emitted from natural sources, e.g. plants or buildings (Suchorab et al. 2017). The anthropological particles mainly originate from combustion sources, such as residential coal-burning and transportation (Połednik 2013a). Suspended particles may comprise various toxic substances, such as heavy metals and polycyclic aromatic hydrocarbons, including benzo(a)pyrene, as well as dioxins and furans (Klejnowski et al. 2010). At present, regular monitoring in cities encompasses only the mass concentrations of PM_{10} and $PM_{2.5}$ particles (characterized by an aerodynamic diameter smaller than 10 and 2.5 µm, respectively). The Air Quality Guidelines (WHO 2006) set the legal standards for PM_{10} and $PM_{2.5}$.

Currently, greater emphasis is placed on the concentrations of ultrafine particles (UFP) having a diameter smaller than 100 nm. These particles constitute a serious health risk (Kumar et al. 2014). Hence, the share of ultrafine particles in the total particle concentration should be taken into account in the urban air quality monitoring. Due to the constant changes of UFP, both of chemical and physical nature, their monitoring is difficult (Sabaliauskas et al. 2012). There is also a multitude of other factors – such as wind direction and strength, temperature inversion and type of urban sprawl – that affect the time and spatial variability of the UFP levels (Kozawa et al. 2012, Sartini et al. 2013). Urban sprawl may reduce the air exchange and form an urban canopy layer (UCL), with thickness dependent on the height of buildings. The aerosol particle concentrations in deep street canyons may exceed the values in other city parts by several times. Since there are no established permissible levels, the UFP concentrations are not regularly monitored in most urban areas.

The relationship between the concentrations of aerosol particles in ambient air and negative health impact, including respiratory and cardiovascular diseases, was reported in epidemiological studies (Valavanidis et al. 2008). The share of urban population exposed to the concentrations of PM_{10} and $PM_{2.5}$ exceeding the daily limit set by WHO for outdoor air pollution amounted to about 61% and 87%, respectively (EEA 2017). In this study, 403,000 premature deaths in the EU (in 2012) were attributed to the exposure to $PM_{2.5}$. In turn, research conducted in London indicated that the exposure of commuters to particle concentrations in urban areas may be about 1.5-fold greater when traveling by bus or by car, comparing to walking along the sidewalks (Kaur et al. 2005). Fruin et al. (2008) reported that roughly 6% of total time spent commuting in a car in the road environment in the USA may contribute to approximately 36% of daily average exposure to particle number concentrations. In turn, Dons et al. (2012) indicated that approximately 3.6% of the daily time spent while commuting by car in Belgium corresponded to about 14.5% of average daily exposure to particle number concentrations. It can be expected that similar results would be obtained for the people living in other countries within the EU (Knibbs et al. 2011). Due to the increase of traveling time over the years, conducting an accurate assessment of the exposure becomes a necessity. Although multiple studies on the commuting exposure in cities were conducted in recent years, it is still difficult to accurately estimate the human exposure to aerosol particles. These difficulties are related to different characteristics of each route, for instance connected with the intensity of traffic, the number of trucks and passenger cars, or the type of fuel. Moreover, routes vary in terms of topography and the built-up area. In addition, the traffic characteristics along one route are variable and constantly affected by traffic volume in different times of the day, weather conditions and other factors. Clearly higher particle concentrations are usually found in traffic intersections (TIs), which constitute pollution hotspots (Goel and Kumar 2015).

The exposure assessments can be performed using the data from mobile and fixed-site measurements, which provide different types of information. Mobile measurements reflect the exposure of commuters to particles (drivers and passengers), whereas fixed-site measurements aid in determining the exposure of pedestrians. Mobile measurements can be performed to quickly collect large quantities of data over extensive areas.

Additionally, both methods of measurement may provide information pertaining to the particle transformation processes (e.g. coagulation, nucleation, deposition and condensation) occurring on the sidewalk and on-road.

Sidewalks constitute an integral part of the road which is used by pedestrians. The functionality of sidewalks includes walking, as well as other physical activities, e.g. running, cycling, etc. Sidewalks are generally found on both sides of roads, they also cross at intersections with heavy traffic. Therefore, the pedestrians using sidewalks are exposed to the traffic-related emissions, including aerosol particles from exhaust sources, i.e. black carbon and gaseous precursors of secondary PM, e.g. NO_x and VOCs (Janssen et al. 2011, Qiu et al. 2019b). Moreover, pedestrians are exposed to the particles originating from non-exhaust sources, including road dust and road surface abrasions or wear of tires and brakes (Kwak et al. 2013, WHO 2005). Epidemiological and toxicological studies indicate that these pollutants have a negative influence on human health and may cause serious diseases, including cardiopulmonary diseases (Valavanidis et al. 2008, WHO 2013). This, in turn, may contribute to increased lung cancer mortality and may adversely affect the central nervous system (Lee et al. 2019, Sram et al. 2017). In terms of the traffic-related emissions, attention has recently been drawn to UFP with an aerodynamic diameter smaller than 100 nm, predominantly as a result of their physicochemical characteristics and properties, such as high pulmonary deposition efficiency, toxicity related to their large surface area, as well as the amount of adsorbed transition metals (Bliss et al. 2018, Rogula-Kozłowska et al. 2019).

The studies conducted by other authors (Mukerjee et al. 2015, Puett et al. 2014) showed that the exposure to traffic-related particles substantially reduces proportionally to the distance from the road. This implies that the doses inhaled by commuters in vehicles are higher than by pedestrians. This was confirmed by the result reported by Cepeda et al. (2017) and de Nazelle et al. (2017). Nevertheless, commuters in vehicles remain in an environment that is, to a certain degree, isolated from the outdoor air by sealed windows; they also inhale filtered air. Hence, pedestrians may actually breathe higher doses of particles, especially on certain sections of the road (such as TIs with heavy traffic) or under certain weather conditions (Moreno et al 2015, Pirjola et al. 2012).

Numerous studies have recently been performed in cities, drawing attention to the exposure of commuters (Kwak et al. 2018, Moreno et al. 2019, Luengo-Oroz and Reis 2019). Moreover, several studies on the exposure of pedestrians have also been conducted (Choi et al. 2018, Qiu et al. 2019a, Rakowska et al. 2014). However, it is difficult to accurately determine the exposure to aerosol particles from the traffic-related emissions. The difficulty predominantly stems from the variability of the obtained measurement results, which are affected by such factors as road characteristics, intensity of traffic, types of vehicles and fuel used. The arrangement of buildings (forming street canyons), topography and the time of the day influence the traffic intensity, whereas the season governs the meteorological conditions. Weather is especially important in Central and Eastern Europe (CEE), where strong winds, low temperatures, heavy rains and snowfalls in winter hamper the vehicular traffic. However, this region has rarely been studied thus far. Moreover, it needs to be emphasized that the exposure to traffic-related vehicles can be influenced by a large share of old vehicles as well as outdated transportation infrastructure (Eurostat 2019, Taczanowski et al. 2018).

4.2   LOCATION OF THE STUDY

Lublin (łac. Lublinum) is a city with county rights, capital of the Lublin voivodeship and Lublin county and the center of Lublin agglomeration. Lublin is located on the northern border of Lublin Upland, in the vicinity of the border between the North European Plain and Polish Uplands, near the East European Plain. It spans an area of 147 km^2 (Statistics Poland 2013). Bystrzyca Valley divides the city into two parts with distinct landscape: the left-bank part with diversified topography, deep valleys and old loess ravines as well as the right-bank part with flatter and less diversified topography. Bystrzyca River has two main tributaries within municipal area: Czerniejówka and Czechówka. Nędznica – also known as Krężniczanka – is another river which flows through the city.

Lublin is located in the humid continental climate zone. The mean annual air temperature ranges from +7.0°C to +8.0°C. July and August are the warmest months, with mean temperature of approximately +19°C, whereas January and February constitute the coldest months, with mean temperature of about −5.0°C. Total annual precipitation is roughly 540 mm. Snow retention ranges from 70 to 90 days (climate-data.org 2020) (Table 4.1).

Lublin is the ninth most populated city in Poland, second in Lesser Poland (339,547 people in 2019) (Statistics Poland 2020a). In terms of area, it is the sixteenth largest city in Poland (147 km^2).

The diagram in Figure 4.1 presents the growth of population throughout the past 400 years. The population of Lublin in 1999 was estimated at 359,154, which the highest number in the history of the city.

Lublin is the largest city of Poland on the eastern bank of the Vistula River. It is approximately 170 km southeast of Warsaw. Lublin is considered as a good place for foreign investments. According to the analytical Financial Times Group, Lublin is one of the highest ranked cities for business in Poland (Kondracki 2002). In turn, the Foreign Direct Investment ranking awarded Lublin the second place among larger cities in Poland in terms of cost-effectiveness. Lublin is notable for a high standard of living as well as extensive green spaces (Lublin City Office 2017). Lublin also has its own airport, located approximately 10 km (6.2 miles) southeast of the city. In 2018, the Lublin Airport had 8 destinations and served more than 450,000, making it the largest

Table 4.1 Climate data for Lublin (1936–2011 (weatheronline.pl)

Month	Jan	Feb	Mar	Apr	May	Jun	Jul	Aug	Sep	Oct	Nov	Dec	Year
Record high °C	18.0	16.2	22.0	27.2	35.7	33.9	35.0	37.0	33.2	25.0	18.9	15.0	37.0
Average high °C	-0.7	0.4	5.7	12.7	18.4	21.4	23.3	23.0	18.1	12.3	5.4	0.8	11.8
Daily mean °C	-3.1	-2.5	1.6	7.8	13.1	16.2	17.9	17.4	12.9	7.9	2.6	-1.4	7.6
Average low °C	-5.9	-5.7	-2	3.0	7.7	10.7	12.5	12.0	8.2	4.0	0.0	-3.9	3.5
Record low °C	-32.2	-31.1	-30.9	-7.2	-4.1	0.0	2.0	0.0	-4	-7.6	-17.9	-23.9	-32.2
Avg. precipitation mm	22.7	25.9	27.3	42.4	51.1	66.6	71.5	64.0	55.5	40.6	36.7	33.6	537.9
Avg. precipitation days	23.3	19.5	18.4	13.1	13.0	11.8	12.3	9.3	11.2	13.3	18.1	20.8	184.1
Avg. RH%	88.7	85.9	79.8	68.9	71.9	73.7	75.1	74.4	79.8	84.0	89.4	90.2	80.1
Mean sunshine hours	53	73	115	174	226	237	238	248	165	124	48	37	1,738

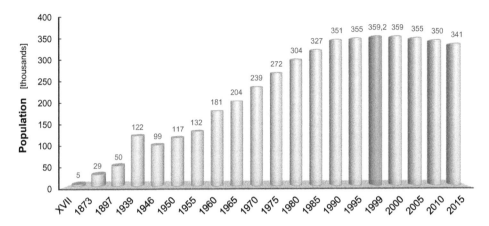

Figure 4.1 Population growth in Lublin (Based on Statistics Poland 2017).

airport in Eastern Poland. The airport has a direct train and bus connection with the city. Lublin is found at the intersections of the S12, S17 and S19 expressways. The transit traffic bypasses the city center via expressways. Moreover, Lublin has trolleybuses, which are found only in four Polish cities.

4.3   CHARACTERIZATION OF TRAFFIC-RELATED POLLUTION IN LUBLIN

The transportation fleet of Lublin predominantly comprises relatively old, gasoline and diesel vehicles, including trucks, vans, cars, buses and electric trolleybuses. Over 50% of all passenger cars are older than 15 years, whereas the public transportation buses are 21 years old, on average (Statistics Poland 2017).

In Lublin, the particle concentrations in air are characterized by seasonal and daily variations (Południk 2013b). These particles are generated from various sources. Lublin is a typical Central and Eastern European city as far as the lay of land, climatic and weather conditions, as well as the transportation structure are concerned. In line with the data provided by Karagulian et al. (2015), the percentage of $PM_{2.5}$ in the total particle emissions is as follows: domestic furnaces 32%, road traffic 19%, industry 17%, natural sources 16% and other anthropogenic sources 16%. In turn, as far as PM_{10} is concerned, the emissions from domestic furnaces are even greater, reaching 45%. The remainder consists of undefined anthropogenic sources 26%, industry 18%, road traffic 8% and natural sources 3%.

The number of vehicles directly affects the concentrations of traffic-related particles. In 2016, there were 217,311 registered vehicles in Lublin, including: 76.9% passenger cars, 13.3% trucks, 5.1% motorcycles, 0.7% municipal transportation vehicles and 4% other vehicles, e.g. tractors, ambulances and special purpose vehicles (BIP 2017).

In terms of PM_{10} emissions, according to a report published by the Voivodship Environmental Protection Inspectorate in Lublin (WIOS 2016), the annual emission of

Table 4.2 Estimated PM_{10} emission from the
Lubelskie Voivodship (WIOS 2018)

Type of emission	Amount Mg/year
Surface	15,798.0
Linear	2,334.1
Point	1,653.1
Agricultural	4,103.9
Unorganized	924.4
Total emission	24,813.5

traffic-related particles between 2013 and 2015 amounted to 2,474.7 Mg/year, i.e. about 10% of the total particle emission from all other sources.

In Lublin, according to the report of Lubelskie Voivodship Environmental Protection Inspectorate *concerning the condition of environment in the Lubelskie Voivodship in 2017* (WIOS 2018), the total annual emission of PM_{10} from linear sources, including roads and other communication pathways, amounted to 2,334.1 Mg/y in 2017, which constituted approximately 9.4% of total emission of these pollutants (Table 4.2). The quantity of linear emission is mainly governed by the pollutants originating from communication pathways.

According to the data by TomTom, a Dutch company that specializes in the creation of navigation maps used in vehicles around the world, Lublin is one of the cities with the highest traffic congestion in Europe. In a ranking presented in 2017, prepared on the basis of multiannual observations and measurements as well as using diverse data sources, e.g. acquired from the application and car navigation system users or reported by municipal services using continuous monitoring systems, Lublin reached the 12th place, right after Rome, overtaking Paris and Brussels (14th and 15th places, respectively). For reference, Warsaw reached the 19th place (TomTom 2017). According to the same source, the residents of Lublin spend on average 37 minutes in traffic jams, which amounts to approximately 140 hours a year. The rush hours last from 7:00 to 20:00 as well as from 3:00 to 4:00 (TomTom 2019).

4.4 NOVEL STUDIES ON TRAFFIC-RELATED ENVIRONMENTAL POLLUTION IN LUBLIN

The first comprehensive studies on the traffic-related environmental pollution were performed in Lublin in 1995 (Pawłowski et al. 1995). This study was conducted on $800 \times 1,000$ m area in the city center, encompassing busy Krakowskie Przedmieście, Kołłątaja, Hempla, Narutowicza and Lipowej streets, as well as the adjacent streets with lower traffic volume (Figure 4.2).

The studies involved 24-hour measurements of car traffic congestion at six crossroads of the busiest streets. The measurements consisted in counting the passing cars, which were allocated to five categories: passenger cars, vans, trucks, buses and other vehicles. While determining the emission of traffic-related pollution, attention was focused on the afternoon rush hour (3:00–4:00). The calculations accounted for the fuel consumption and pollution emission indices for particular categories of vehicles. It was

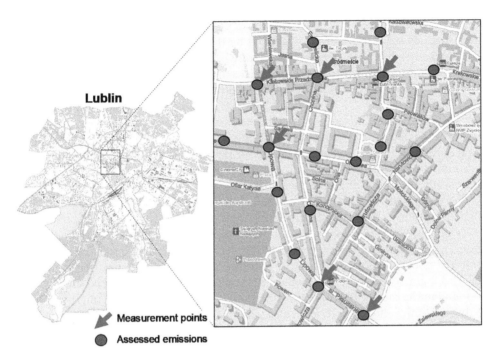

Measurement points

Assessed emissions

Figure 4.2 Plan of the investigated area.

assumed that each lane constitutes an independent emission source. The calculations accounted for all relevant meteorological parameters, i.e. wind direction, wind speed, atmospheric stability classes (according to Pasquill), air temperature and height of the mixing stratum. On the basis of the data collected for the selected street segments, the emissions of basic pollutants, i.e. carbon monoxide, nitrogen oxides, hydrocarbons and sulfur dioxide, were determined. The calculations were performed for the investigated area divided by a rectangular grid with the dimensions of 50×50 m. Receptors were placed in each grid node at a height of 1 m. Moreover, the arrangement of streets in the analyzed area was divided into 20 straight segments with constant vehicle traffic intensity, which constituted separate linear sources. The location of particular segments was determined by superimposing them on the coordinate system and indicating their beginnings and ends. The determined values of momentary (30-minute) concentrations enabled drawing a concentration isoline of these pollutants with the interpolation method. The obtained results enabled to state that the spatial distribution of emission of all analyzed pollutants is concentrated in the direct adjacency of the street. The highest horizontal concentration gradients are found there, which gradually decrease as the distance from the street increase. The direct impact of traffic on the pollutant concentrations was observed in a band up to 100 from the street axis. Certain deviations were noted in the case of high-density housing development along the street as well as in the places at a certain distance from the streets, where relatively high concentrations of the considered pollutants were observed. It was suggested that it may result from the effect of emissions from adjacent streets or due to the influence of a relatively intense linear source on its extension in the axial direction. Elevated pollutant concentrations on the extensions of some streets, forming so-called street canyons, which could have been influenced by the winds blowing in the axial direction as well as high housing development, were provided as an example. Additionally, exemplary influence of parallel streets and accumulation of pollutants emitted from both streets were provided. The calculations omitted the pollution from other, non-traffic-related sources, which constituted the background. This enabled to determine the share of traffic-related emissions in the total balance of air pollution in Lublin. The analysis of the obtained results confirmed the significant influence of this emission on the exceeded permissible concentrations of air pollutants.

Figure 4.3 presents the spatial distribution of momentary CO, NO_2, HC (hydrocarbons) and SO_2 in the analyzed area in Lublin.

The estimated maximum CO concentration reached the level of 16.38 mg/m^3, which constitutes over 300% of permissible concentration (5 mg/m^3). CO concentration exceeding 10 mg/m^3 was confirmed in several locations within the examined area. The analysis of emissions and spread of NO_2 exhibited a similar trend. The obtained maximum concentration (1.923 mg/m^3) was almost four-fold higher than the limit value (0.5 mg/m^3) and three-fold higher in some places. A similar spatial distribution was found in the case of HC, with the maximum concentration of 1.437 mg/m^3 and exceeding the limit values in several locations. It was also mentioned that prolonged high concentration of hydrocarbon under the conditions of high insulation may constitute a source of aldehydes and ozone, contributing to the occurrence of so-called photochemical smog. SO_2 constituted the only analyzed pollutant for which no exceeding of the permissible values was noted (0.6 mg/m^3). The spatial distribution of sulfur dioxide emission was similar to that for other pollutants. The maximum concentration in the analyzed area amounted to 0.15 mg/m^3, whereas the remaining values of this pollutant were much higher. It was observed that such concentrations of SO_2 can be connected with relatively low sulfur content in car fuels.

The research concluded that the actual concentrations in streets can be much higher than those obtained by means of computer simulations. The cause might be unstable traffic intensity connected with the operation of traffic lights. Moreover, the concentrations of pollutants at crossroads are usually higher due to greater traffic congestion as well as increased emissions of these pollutants as a result of increased engine load during the moving off.

4.5   MONITORING OF TRAFFIC-RELATED POLLUTION PERFORMED BY VOIVODSHIP ENVIRONMENTAL PROTECTION INSPECTORATE

In Lublin, similarly to other Polish cities, the concentrations of UFP are not monitored. Only the concentrations of PM$_{10}$ and PM$_{2.5}$ are monitored by Voivodship Environmental Protection Inspectorate in a monitoring station at Obywatelska street. The results are available on-line as mean hourly values. The concentrations of PM$_{10}$ have been continuously monitored since 2006, whereas PM$_{2.5}$ has been monitored since 2009 (WIOS 2017).

Only particle mass concentrations are measured in Lublin, like in many other urban agglomerations around the world. The mean hourly value of PM$_{2.5}$ concentrations

Figure 4.3 Differences in momentary traffic-related air pollution (CO, NO$_2$, HC and SO$_2$) in the center of Lublin (Pawłowski et al. 1995, modified).

from the entire measurement period equals 30.0 µg/m^3 with a standard deviation of 30.5 µg/m^3. Concentrations were higher than 25 µg/m^3 for 42.7% of the measurement time, whereas concentrations greater than 200 µg/m^3 occurred for 0.5% of the measurement time. The maximum value was reached on 6 December 2012 at 10 p.m., reaching 574.9 200 µg/m^3. In turn, for the PM$_{10}$ concentrations from the entire measurement period equaled 31.3 µg/m^3 with a standard deviation of 28.9 µg/m^3. The concentrations were higher than 50 µg/m³ for 13.72% of the measurement time, whereas concentrations greater than 200 µg/m³ occurred for 0.43% of the measurement time. The maximum concentration was observed on 9th January 2016 at 11:00 and reached 551.1 µg/m³. The measurements conducted over many years indicated no increasing or decreasing tendency in terms of the PM$_{2.5}$ and PM$_{10}$ concentrations. Although the number of vehicles registered in the Municipality of Lublin increased substantially, the values of pollutants remain similar. Nevertheless, this does not mean that motor vehicles have no material impact on the levels of suspended particles. New cars, equipped with improved engines that meet the EU emission standards, as well as hybrid and electric cars, replace the old passenger cars on a regular basis.

The results of multiannual monitoring of aerosol particle concentrations in Lublin show that the main cause for their increase involves the combustion processes for heating buildings and flats. The former are usually performed in domestic furnaces.

During the heating period, which in Lublin usually lasts from October to March (often until mid-April), there are noticeable increases in the concentrations of PM$_{2.5}$ and PM$_{10}$, the maximum values of which exceed 500 µg/m³. In addition, the standard deviations of the measured concentrations are observed. This is probably connected with the periodic character (usually daily) combustion process in domestic furnaces, which are prevalent in the urban and suburban areas of Lublin.

4.6 INVESTIGATION OF ELEMENTAL COMPOSITION OF TRAFFIC-RELATED AEROSOL POLLUTANTS

Determination of metals in sedimented and deposited aerosol particles was performed using inductively coupled plasma mass spectrometry (ICP-MS) apparatus. The samples for investigation were collected from various locations of the city, which were characterized by diversified traffic intensity (Figure 4.4). After sieving, the samples were mineralized and the elements were determined quantitatively using model-based calibration curves. The results of determinations are presented in Figure 4.5.

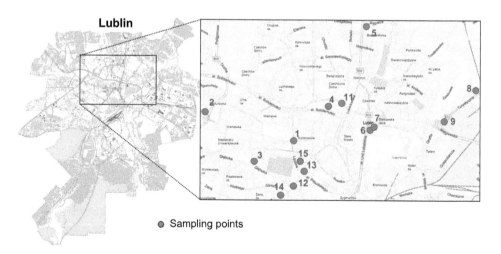

Figure 4.4 Map of Lublin with marked sample collection locations.

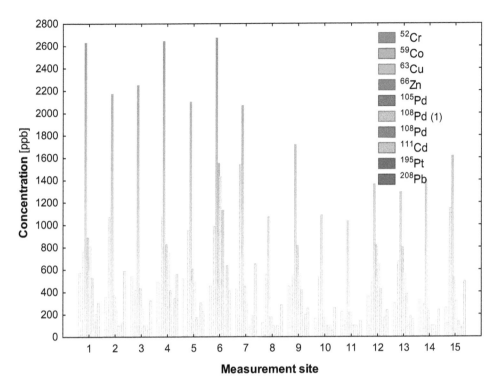

Figure 4.5 Concentrations of metals in the samples of particles deposited in different locations of Lublin.

The performed studies indicated that in the particles deposited in all considered locations of the city, Zn occurred in the highest amounts (Czerwiński 2019). Higher concentrations of palladium isotopes as well as cadmium, platinum and lead occurred in the samples collected from busy streets. The obtained results confirmed the expected dependence of metal concentrations in the deposited particles on the intensity of vehicle traffic and the emitted amount of related pollution. Galvanized car bodies as well as catalytic converters could have had a significant influence on the results.

4.7   STUDIES ON THE CONCENTRATIONS OF AEROSOL PARTICLES IN DIFFERENT LOCATIONS OF THE CITY

The particle number and particle mass concentrations were measured in different locations of Lublin in October of 2017 (Test 1) as well as March (Test 2) and April of 2018 (Test 3).

The measurements were performed by means of a Mobile Air Pollution Analytic Laboratory (MAPAL) installed in a Renault Kangoo (year of registration: 2008; displacement: 1461 cc) that was driven on the planned route. Stops were made at fixed measurement points on the route during which 10-minute fixed-site measurements were performed (Figure 4.6).

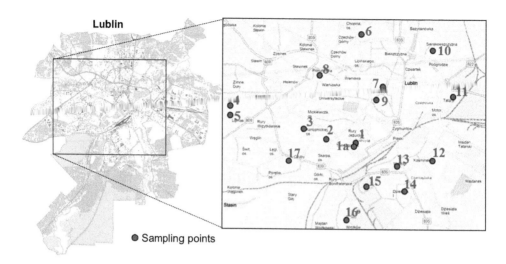

Figure 4.6 Map of Lublin with marked sampling locations of aerosol particles.

The locations in which fixed-site measurements were performed were scattered in different districts of the city. These locations also differed in terms of traffic intensity – from several to several thousand passing cars per hour. Tests 1 and 2 involved three measurement runs each (a, b, c), whereas Test 3 comprised a single measurement run (a).

MAPAL comprised the Grimm Aerosol Spectrometer 1.109 with Nano Sizer 1.321 (Grimm Aerosol, Germany) system, enabling to remove humidity from samples and conduct real-time measurements related to number concentrations of particles (10 nm up to 32 µm) in different size channels as well as inhalable, thoracic and alveoli particle mass fractions in line with the EN 481 standard. The number concentrations of particles sized from 0.02 to approximately 1 µm (PN_1) were measured with P-Trak model 8525 ultrafine particle counter (TSI Inc., USA). In turn, OPS 3330 optical spectrometer (TSI Inc., USA) was used to measure the number concentrations and size distribution of particles larger than 0.3 µm ($PN_{0.3-0.5}$, $PN_{0.5-1}$, PN_{1-2}, PN_{2-5}, PN_{5-10} and $PN_{>10}$). DustTrak DRX model 8533 aerosol monitor (TSI, Inc., USA) was used to measure particle mass concentrations and obtain their approximations for PM_1, $PM_{2.5}$, RESP, PM_{10} and TSP (particles with an aerodynamic diameter equal to or less than 1, 2.5, 4, 10 µm and total suspended particles, respectively). DustTrak monitor was calibrated using the standard real-time size correction factor. The approximations obtained for the particle mass concentration values are not actual gravimetric values. All the Dust-Trak results presented in this paper omit the term "approximation" to avoid confusion. The MAPAL instruments were supplied with air via tubes. The endpoints of these tubes were located in the middle of the car, on the left, at the height of about 1.7 m. The instruments were calibrated by their respective manufacturers when the measurement campaign started. The instruments were set to 6-second logging interval. Garmin Nuvi 2460LMT GPS (Global Positioning System) device continuously recorded the speed and position of MAPAL. The traffic flow data at the time of measurements were collected by means of a HD 1080P Wide-angle 170° camera installed on the dashboard. The LB-520 thermo-hygrometer (LAB-EL, Poland) was used to measure air temperature and relative humidity inside and outside the vehicle. In this article, only the data related to ultrafine particle number concentrations ($PN_{0.1}$) measured by means of the Grimm Aerosol Spectrometer, submicron particle number concentrations (PN_1) determined with P-Trak, and mass concentrations of fine ($PM_{2.5}$) and coarse (PM_{10}) particles measured by using DustTrak DRX are present. At the beginning of the experiment, the timestamps of all the instruments were matched.

Some results related to the mobile and fixed-site measurements of particle concentrations are presented in Table 4.3 and Figures 4.7–4.9.

The presented results indicate the temporal and spatial variability of particle number and mass concentrations in the considered area of Lublin. Mobile and fixed-site measurements performed in test runs show the existence of hot spots, in which relatively high concentrations of aerosol particles, mainly traffic-related, are found. These are the main intersections of the busiest streets and their sections where traffic jams occur. In these places, cars brake more often, move slowly, or rapidly accelerate, simultaneously emitting increased amounts of air pollutants. In the considered points, the topography as well as dense and relatively high urban development on both sides of the street are conducive to the accumulation of air pollutants. Such development leads to the creation of street canyons, where substantial amounts of traffic-related pollutants are accumulated. In intersections and street canyons, drivers and passengers are exposed to increased concentrations of harmful pollutants. These pollutants

Table 4.3 Mean concentrations of PN_1 aerosol particles obtained in particular measurement points in the course of test runs (a), (b) and (c) during Test 1, Test 2 and Test 3 (data obtained by means of the P-Trak instrument)

Test 1

Run	Point	PN1 pt/cm³
a	1	4009
	2	4239
	3	6062
	4	6741
	5	6142
	6	9422
	1	4469
b	1	24642
	7	28565
	10	21595
	11	22030
	12	20913
c	1	6207
	9	8502
	8	13147
	17	7220
	15	5219
	13	4698
	14	4181
	16	4581

Test 2

Run	Point	PN1 pt/cm³
a	1	2148
	2	2171
	3	3780
	4	2029
	5	1777
	6	4922
	7	5768
	8	7002
b	9	3267
	10	3993
	11	4382
	12	5493
	13	8849
	14	11273
	15	4609
	16	3181
c	1	4050
	2	4814
	3	5737
	4	3271
	5	4386
	8	6613
	6	4870

Test 3

Run	Point	PN1 pt/cm³
a	1	2223
	2	3226
	3	3438
	4	2560
	5	3193
	8	6888
	6	2955
	10	2341
	7	3601
	9	4246

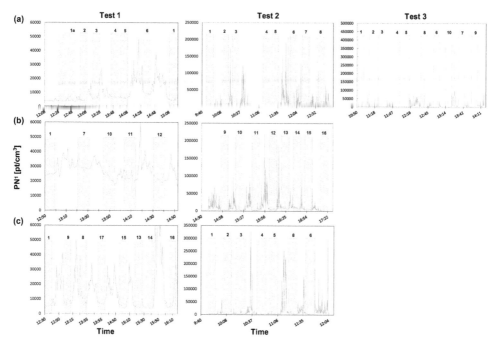

Figure 4.7 Particle number concentrations (PN$_I$) acquired from mobile and fixed-site measurements in the course of test runs (a), (b) and (c) during Test I, Test 2 and Test 3 (data obtained by means of the P-Trak instrument).

can – to a certain extent – be stopped by relatively tight car glass. Air filtering and purifying devices installed in cars can also be helpful. Pedestrians and people walking, jogging, or physically exercising in the vicinity are in a much worse situation. Usually, increased breathing frequency of such people can lead to the inhalation of extremely high amounts of substances polluting urban air.

4.8   INVESTIGATIONS OF THE CONCENTRATION OF AEROSOL PARTICLES AND EXPOSURE ALONG A BUSY STREET

During the research, mobile and fixed-site measurements were conducted along a 2.1 km long route, constituting part of one of the busiest streets in Lublin, with 3- or 4-story buildings found on each side. The considered four-lane street comprises three 4-way TIs with traffic lights. The traffic pattern is similar to a typical bimodal pattern found in the majority of urban areas, reaching maximum in the morning (7:00–8:00) and afternoon (3:00–4:00) peak hours. During the rush hours, the volume of traffic at each of the considered intersections may exceed 2,000 vehicles per hour (the maximum vehicle number reported is 3,000) (RBA 2018). In summer and during the off-peak hours, the traffic is significantly less intense (Figure 4.10).

The MAPAL vehicle was used to perform measurements. The vehicle stopped at 12 measurement points along the route, with 5-minute fixed-site measurements performed at each point. During the measurements, MAPAL was parked on the sidewalk.

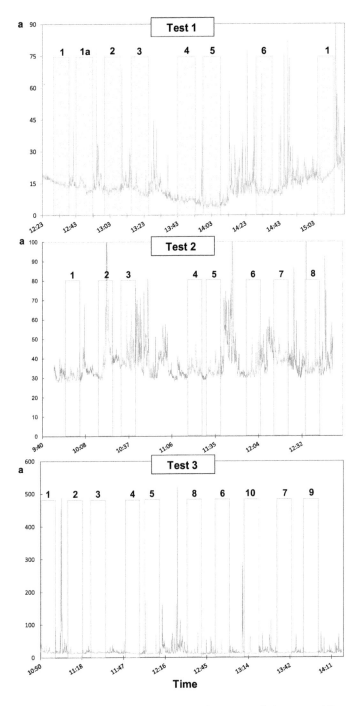

Figure 4.8 Particle mass concentrations (PM$_{2.5}$) acquired from mobile and fixed-site measurements in the course of test run (a) during Test 1, Test 2 and Test 3 (data obtained by means of the DustTrak instrument).

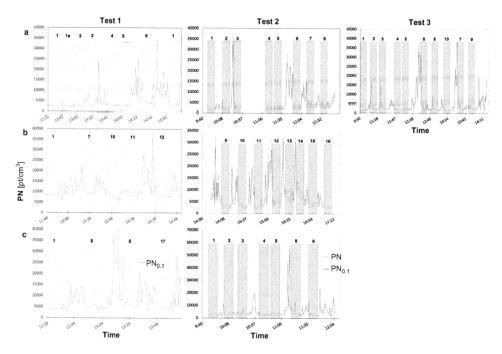

Figure 4.9 Particle number concentrations of PN obtained from mobile and fixed-site measurements in the course of test runs (a), (b) and (c) during Test 1, Test 2 and Test 3 (data obtained by means of the Grimm instrument).

The test runs were carried out in both directions (point 0 to point 11 and back). Along the route, some points were omitted due to lack of available parking space. Hence, attempts were made to conduct the measurements in the omitted points during the return trip; however, they were not successful in each case. Six runs were performed each day, in order to account for the morning and evening peak hours, lighter traffic at night, as well as less intense traffic at midday. The monitoring campaigns were carried out over several consecutive days in the spring, summer and fall of 2017 as well as in the winter of 2018. In Lublin, the mean ambient temperatures during those seasons amounted to 9.4°C, 19.1°C, 9.5°C and –1.6°C, respectively (https://www.weatheronline.com). Due to an exceptionally warm December, the average temperature in winter was relatively high; nevertheless, the average temperature dropped well below zero on some days.

The sampling of ambient air PM was carried out along the determined route, with runs in both directions. The fixed-site measurements were performed in 11 sampling points located on the sidewalk. Six points were found on the sidewalk in the section in the vicinity of 4-way TIs (RS-I) with intense traffic, whereas the remaining five points were found away from four-way TIs, where the traffic was less intense (RS-II). Measurements were performed for 5 minutes at each point from the MAPAL vehicle that was parked on the sidewalk. The background concentrations of particles were measured about 350 m away from the investigated street, in a certain distance from other streets, and in the vicinity of a river where the motor vehicle traffic was almost completely absent. Thus, the background levels were affected by traffic-related pollution

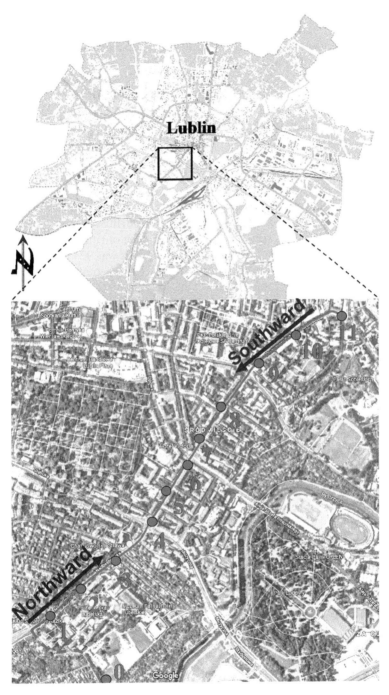

Figure 4.10 Map of Lublin and Google Earth satellite image showing the marked route for mobile monitoring and the location of the fixed-site measurement points on the sidewalk.

to a limited degree or not at all. At the beginning and after the end of each run, the average particle concentrations were measured over 5 minutes and then used as the background for each run. The obtained concentrations of traffic-related particles from each point were calculated as differences of the total size-fractionated particle concentrations and the respective background concentrations.

A total of 6 runs were carried out per test day at fixed times in 4-hour intervals. The measurements were designed in order to take the morning and afternoon peak hours, as well as the off-peak hours in the evening and at night, into account. A full single run lasted for about 95 minutes, on average. However, the exact time was dependent on the traffic intensity and the time of the day. The presented work analyzed the results of measurements acquired over the course of six consecutive runs on a single day that was representative for the particular season. The runs were carried out under relatively stable weather conditions, characterized by differences in natural temperature and relative humidity resulting from the time of the day. In the study, it was assumed that the traffic conditions were the same during each run. The time of the day and the related traffic intensity affected the duration of runs. The runs during the peak hours lasted for 103 ±12 minutes, whereas at night it was 92±8 minutes.

The exposure to particles was determined taking into account the respiratory dose (RD) of inhaled particles obtained from a simplified methodology presented by Qiu et al. (2019a) as well as the respiratory deposition rate (DR) – with the methodology by Joodatnia et al. (2013). Estimation of the total dose of particles inhaled by drivers and passengers (commuters) on the road as well as pedestrians on the sidewalk was performed using equation (4.1):

$$RD = V_T fPCt \tag{4.1}$$

where V_T is the tidal volume, f is the breathing frequency, PC is the averaged particle number or mass concentration and t is the duration of exposure.

The tidal volume and breathing rate are governed by age, gender and the level of activity. The assumed values were $800\,cm^3$ per breath as well as 18 and 21 breaths per minute for a male adult car driver and passenger on the road and for a male adult pedestrian on the sidewalk, respectively. In the case of commuters and pedestrians, an unlimited exposure to traffic-related pollutants was assumed. In turn, in the case of car users, unlimited exposure occurs when the windows of their cars are fully open. It can be assumed that the same situation occurred in the course of the on-road measurements described in this article.

The particle concentrations as well as relative commuter and pedestrian exposure, which were grouped depending on the particle size, part of the route, time of the day, and season, were characterized using descriptive statistics.

4.8.1   The results of spring measurements

The measurements performed along the determined route and in the established measurement points on the sidewalk at different times of the day enabled to track changes in the number and mass concentrations of particles in different size ranges. The time series of the number concentrations of particles within the 20–1,000 nm size range (PN_1) is presented in Figure 4.11, ultrafine particle number concentrations ($PN_{0.1}$) are

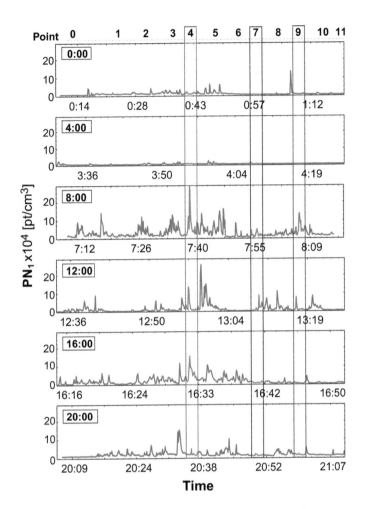

Figure 4.11 Fluctuations of PN$_1$ particle number concentrations in the course of mobile monitoring with indicated part of the route characterized by heaviest traffic and fixed-site measurement points in 6 consecutive measurement periods (0:00, 4:00, 8:00, 12:00, 16:00 and 20:00) (data obtained by means of the P-Trak instrument).

shown in Figure 4.12, and mass concentrations of particles ≤10 μm (PM$_{10}$) are presented in Figure 4.13. The data correspond to mobile and fixed-site measurements conducted in six measurement periods, i.e. at 0:00, 4:00, 8:00, 12:00, 16:00 and 20.00. The data in these figures was obtained using P-Trak, Grimm and DustTrak instruments, respectively.

In spring, the greatest average concentrations were observed in peak traffic hours for the part of route characterized by most intense traffic and amounted to $25.4 \pm 11 \times 10^3$ pt/cm^3 for ultrafine particle number PN$_{0.1}$ (mean ± standard deviation), $29.2 \pm 12 \times 10^3$ pt/cm^3 for total particle number PN, 29.1 ± 7.6 μg/m^3 for mass concentrations of PM$_{2.5}$, as well as 45.4 ± 10.3 μg/m^3 for PM$_{10}$. Depending on the particle size,

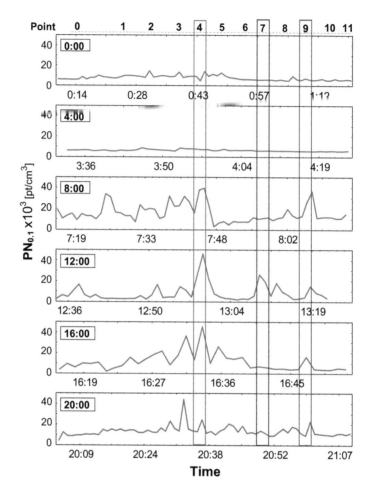

Figure 4.12 Fluctuations of $PN_{0.1}$ particle number concentrations in the course of mobile monitoring with indicated part of the route characterized by heaviest traffic and fixed-site measurement points in 6 consecutive measurement periods (0:00, 4:00, 8:00, 12:00, 16:00 and 20:00) (data obtained by means of the Grimm instrument).

the average particle number concentrations along the entire route and its considered part in peak times were approximately 3–4-fold greater than in the off-peak times. In turn, the value of average particle mass concentrations was roughly twice as high. Moreover, the average values of the considered particle number and mass concentrations as well as percentage of ultrafine particles were lower for the fixed-site measurements, than for the on-road measurements. It was found that the number and mass of particles, regardless of their size range, are deposited in respiratory tract of commuters in greater amounts during the peak hours, compared to the off-peak hours. The average doses of particle inhaled by commuters during the peak periods amounted to $4.8 \pm 2.4 \times 10^9$ pt/h or 29.6 ± 10.7 µg/h (PM_{10}). Moreover, higher particle doses were

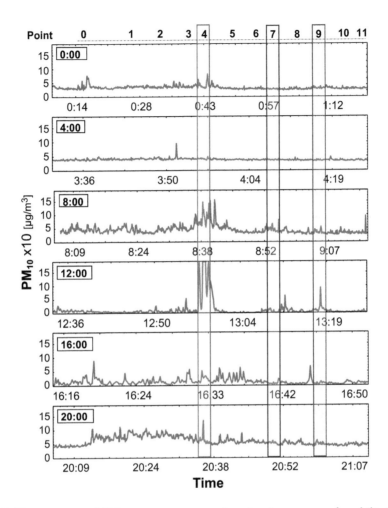

Figure 4.13 Fluctuations of PM_{10} mass concentrations in the course of mobile monitoring with indicated part of the route characterized by heaviest traffic and fixed-site measurement points in 6 consecutive measurement periods (0:00, 4:00, 8:00, 12:00, 4:00 and 8:00) (data obtained by means of the DustTrak instrument).

observed in the considered part of the route with the most intensive traffic, both during the peak and off-peak hours.

The part of the route characterized by the most intense traffic, corresponding to the location of the fixed-site measurement points 4, 7 and 9, are shown in Figures 4.11–4.13. The measurement points 4 and 7 were located in the vicinity of 4-way TIs. The obtained values were varied significantly in the particular measurement periods. The most distinct changes related to the number and mass particle concentrations corresponded to the peak hours, especially in the vicinity of 4-way TIs. Elevated PM_{10} concentrations along the entire route occurred in the evening (at 20:00). Kumar and Goel (2016) reported a similar tendency. Generally, their investigation indicated that coarse particles mainly include non-exhaust sources, i.e. road abrasion and brake and tire wear; in turn, the fine particles occur predominantly due to combustion of fuel in engines. According to the authors, on-road concentrations of $PM_{2.5}$ in the evening were twice the value measured in the mornings, reaching 33 ± 9 and 16 ± 4 $\mu g/m^3$, respectively. Such differences did not occur for coarse particles. In their study, Kuhlbusch et al. (1998) attributed this phenomenon to higher fugitive dust emissions that occur in the evening. These emissions result from higher surface temperature of the road as well as lower surface moisture content. This confirms that coarse on-road particles are predominantly influenced by the emissions from non-exhaust sources, emitted throughout the day. Therefore, factors like different weather conditions during a day, involving variable wind speed and height of the mixing layer (Oleniacz et al. 2016, Zhao et al. 2017) or diversified traffic intensity (Amato et al. 2013) cannot be ignored during investigations.

The concentrations of particular particle types differed when measured with various instruments. The average mass concentrations of PM_1, $PM_{2.5}$ and PM_{10} measured in the course of mobile monitoring and in fixed measurement points using the Grimm instrument were 2.1-, 1.7- and 1.3-fold lower, respectively, than the values obtained by means of DustTrak. Moreover, the particle number concentrations indicated using Grimm were lower as well, e.g. reaching 1.8-fold lower value of PN_1 than indicated by P-Trak. Additionally, the results obtained by means of Grimm significantly differed from those obtained using other TSI instruments. When conducting the mass particle concentration measurements, the settings of calibration factors affecting the accuracy of measurements in comparison to the reference gravimetric method may be highly important for measuring particle mass concentrations.

The discrepancies between the particle concentrations measured using the TSI instruments and Grimm that increased as the particle size decreased may also stem from the calibration or other factors related to instruments. This tendency was reported by Cheng (2008), who monitored the PM concentrations in an iron foundry using TSI and GRIMM instruments, as well as by Polednik (2021), who conducted measurements in a dental office. Additionally, the traffic-related aerosols can be characterized by diversified composition, which may be vastly different from the real-time instruments employed for calibration. Dispersed fluids, including suspended water droplets, can be measured as airborne solid particles potentially distorting the obtained results. For the sake of simplification, the following part of this paper is focused on the results collected by means of the Grimm instrument.

Table 4.4 shows the basic statistical information pertaining to particle number ($PN_{0.1}$, PN) and mass concentrations ($PM_{2.5}$, PM_{10}) for the performed monitoring runs along the entire and part of route where the traffic was characterized with greatest intensity (between two 4-way TIs) during peak (8:00 and 16:00) and off-peak traffic hours (0:00 and 4:00). On the other hand, Table 4.3 shows the data for particular fixed-site measurement points, also for the peak and off-peak hours. On average, the particle number concentrations along the entire and part of route where the traffic was characterized with greatest intensity were approximately from 3- to 4-fold greater than during the off-peak hours, depending on particle size. In turn, the particle mass concentrations were, on average, about twice higher, Moreover, the average values of the obtained particle mass and number concentrations were greater in the case of the on-road measurements than for those performed in fixed-site points.

Table 4.4 Statistics of the particle concentrations on the investigated road in particular hours (measured along the route without stopping) (driver)

Run	$PN_{0.1} \times 10^3$ pt/cm^3	$PN \times 10^3$ pt/cm^3	$PM_{2.5}$ µg/m^3	PM_{10} µg/m^3
0:00	6.6(2.4)5.8/4.2–12.8	7.9(2.7)7.1/4.9–14.6	11.7(2.6)10.7/7.9–21.5	16(3.9)14.9/11.7–26.2
4:00	6.0(2.0)5.3/4.2–11.8	7.4(2.2)6.3/5.6–13.1	11.9(2.4)11.1/8.4–18.3	15.5(4)15/10.4–26.2
8:00	14.2(7.3)11.8/2.2–31.1	16.9(7.9)14.9/3–34.3	21.2(6.4)21.6/8.1–36.9	34.8(14)33.9/11.4–87.6
12:00	6.6(4.8)4.7/2.5–18.4	7.5(5.0)5.7/3.1–19.2	8.9(3.5)7.7/4.6–17.6	18.2(11.4)16.4/7.8–54
16:00	10.5(6.2)8.6/2.2–24.7	11.9(6.7)10/2.9–26.1	11.9(5.4)10.8/5.4–31.6	17.2(6.9)16.3/8.8–40.4
20:00	10.2(3.6)9.4/4.2–20.8	12.5(4.1)11.8/6.1–22.8	20.2(6.2)18.4/11.6–34.9	27.5(9.5)25.5/13.3–48.1

Arithmetic average(SD)median/min-max.

Table 4.4 shows the data pertaining to the exposure for commuters. The greatest average concentrations of $PN_{0.1}$ ($14.2 \pm 7.3 \times 10^3$ pt/cm³; mean $\pm$ standard deviation), PN ($16.9 \pm 7.9 \times 10^3$ pt/cm³), as well as $PM_{2.5}$ (21.2 ± 6.4 µg/m³) and PM_{10} (34.8 ± 14 µg/m³) were measured during the morning rush hour (at 8:00). In turn, the lowest particle concentrations, characterized by smallest variations, were obtained at night (0:00 and 4:00).

In the case of fixed-site measurement points (Table 4.5), the points near 4-way TIs were characterized by greatest average number and mass particle concentrations. Regardless of the hour of measurement (peak or off-peak), the greatest average number and mass particle concentrations were observed at points 5 and 6, respectively. However, the values from peak hours were multiple times higher than those from off-peak hours. For instance, during peak hours, the greatest reported average concentrations of $PN_{0.1}$ and total PN particles at point 5 reached 33.6×10^3 and 37.6×10^3 pt/cm³, respectively. In the case of the off-hours, these values equaled 5.7×10^3 and 7.1×10^3 pt/cm³, respectively. The average mass concentrations of $PM_{2.5}$ and PM_{10} in point 5 for peak hours amounted to 33.8 and 53.7 µg/m³, respectively, whereas for the off-peak hours, it was 13.0 and 17.5 µg/m³. In turn, the greatest average mass concentrations of $PM_{2.5}$ and PM_{10} were observed in the off-peak hours in point 6, reaching 15.0 and 18.8 µg/m³, respectively.

Figure 4.14 shows the changes in particle concentrations within the investigated measurement points. The changes in the average number concentrations of $PN_{0.1}$, total particles PN, as well as $PM_{2.5}$ and PM_{10} in particular measurement points during the considered periods confirm that the exposure of pedestrians to particles is dependent upon the time of the day as well as the location along the route.

The percentage of $PN_{0.1}$ number concentrations in total PN particle number concentrations in the course of mobile and fixed-site measurements along the entire monitoring route or its parts (between two 4-way TIs – i.e. points 4, 5 and 6) is shown in Table 4.6.

It can be seen that a greater share of ultrafine particles was found in the case of mobile monitoring, as opposed to fixed-site measurement points. The greatest percentage of ultrafine particles in the total particle number concentrations was reported in the course mobile monitoring route, as opposed to the fixed measurement points located on the adjacent sidewalk. Moreover, these percentages were greater in the morning and afternoon peak hours than at night. According to literature reports (Joodatnia et al. 2013, Goel & Kumar 2015), this results from a higher number of vehicles, which contributes to elevated concentration of the generated nucleation mode particles (size range 5–30 nm).

The estimated doses of the investigated particles inhaled by pedestrians and commuters following an hour on the studied route in both peak and off-peak hours are shown in Table 4.7. The presented values were acquired using the Grimm instrument. The results obtained from TSI instruments would be about twice higher. The data show that commuters are exposed to greater particle doses at peak hours, reaching on average $4.8 \pm 2.4 \times 10^9$ pt/h or 29.6 ± 10.7 µg/h (PM_{10}). When compared to the values estimated using TSI instruments by Połednik (2013c), they were approximately 7- and 5-fold greater than those inhaled within a certain distance from busy streets (1.4×10^9 pt/h for PN and 12.5 µg/h of PM_{10}). According to Joodatnia et al. (2013), the average dose of PN inhaled by commuters in Guildford (UK) was slightly greater (5.5×10^8 pt/min). This might be related to the application of DMS50 instrument, which measures particles in the size range of 5–560 nm. Furthermore, such differences are to be expected, considering all the factors affecting the dose estimation.

Table 4.5 Statistics of the particle concentrations in fixed measurement points (stops were made along the route) (pedestrian)

Point	$PN_{0.1} \times 10^3$ pt/cm³	$PN \times 10^3$ pt/cm³	$PM_{2.5}$ μg/m³	PM_{10} μg/m³
0	5.3(2.4)5.1/2.0–10.2	6.5(2.9)6.3/2.5–11.7	11.5(4.7)12/4.5–18.3	18.9(8.4)17/8.0–34.8
1	9.8(5.4)8.1/3.1–22.7	11.8(6.3)10.0/3.9–30.4	16.7(8.2)13.6/7.5–44	26.1(12.3)19.6/11–50.5
2	6.8(3)7.4/2.3–12.1	8.4(3.8)8.9/2.8–14.9	14.4(7.5)12.5/5.4–31.2	21.1(9.1)16.7/9.5–40.9
3	9.3(7.4)7.2/2.2–38.2	10.9(8.4)8.5/2.5–44.3	14.8(9)12.5/5.6–47.4	22.9(10.1)19.8/9.3–55.7
4	11.5(8.9)8.5/1.8–42.8	13.5(9.5)10.5/2.1–46.0	27(44.7)15.8/5.8–247	62.2(156.8)22.8/14–843
5	13.8(13.5)7.1/4.7–43.5	15.9(14.6)8.5/6.2–47.2	18.7(9.5)14.4/10.8–40	27.3(16.3)19.1/13.1–62.3
6	15.4(11.3)10.1/4.5–42.3	18.2(12.5)13.1/5.6–45.4	19.9(9)16.9/7.3–39.5	24.8(10.5)24/9.2–44
7	5.7(2.3)5.3/2.4–15.6	7.0(2.6)6.5/2.8–17.9	11.2(1.9)10.8/4.8–15.1	14.5(3)14.5/6.5–19.9
8	9.4(6.6)7.9/5.3–34.9	10.9(6.6)9.6/6.5–36.2	14.3(2.6)14/10.9–19.4	20.8(7.9)18.2/12.4–36.1
9	7.1(3.4)5.6/2.6–15.9	8.6(4.0)6.8/3.4–17.9	13.4(5.4)10.7/6.6–26.5	19.5(8.4)15.1/11–42.4
10	8.1(6.5)5.1/2.8–31.9	9.8(7.6)6.1/3.6–36.7	14.3(7.6)10.6/5.7–33.6	18.9(10)14.6/7.1–44.2
11	6.6(2.6)4.9/4.1–12.9	8.0(3.1)6.0/4.9–15.3	12.7(4.8)9.9/6.9–20.7	17.5(8.1)13.1/8.4–33.7

Arithmetic average(SD)median/min-max.

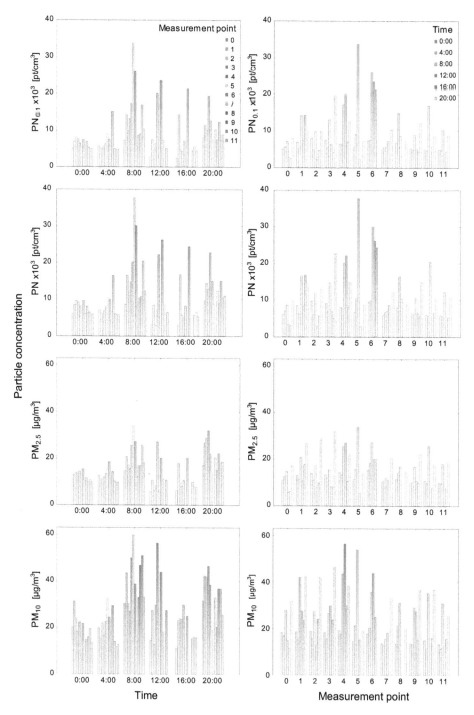

Figure 4.14 Mean concentrations of $PN_{0.1}$, PN, $PM_{2.5}$ and PM_{10} in the considered measurementperiodsandinparticularfixed-sitemeasurementpoints(dataobtainedbymeansof the Grimm instrument).

Table 4.6 Percentage (%) of $PN_{0.1}$ concentration in total number concentration (PN)

Type of measurement	Monitoring period					
	0:00	4:00	8:00	12:00	16:00	20:00
Route without measuring points	82.9	81.7	85.6	86.8	86.1	81.4
All measurement points	81.3	80.8	84.0	84.3	83.1	81.2

The details pertaining to the particle exposure as well as the estimated doses of particles inhaled by commuters and pedestrians following an hour spent on the investigated route or the parts with the most intense traffic are shown in Figure 4.15. It was observed that higher doses of particle are received by those people in the part with intense traffic, as compared to the entire route. It seems unusual that pedestrians inhale greater particle doses (except for $PM_{2.5}$) than commuters. However, this is influenced by the different breathing frequency of pedestrians and commuters. In contrast to the results obtained by Kaur et al. (2005), the findings from this study show that pedestrians are exposed to traffic-related pollutants to a greater extent than commuters, especially since the latter mostly drive with closed windows and breathe filtered air.

The conducted measurements of ambient air quality and evaluated particle exposure could be significantly affected by the weather conditions. In the course of the measurement period, the wind was predominantly blowing from south-west, while its speed was in the range from 1.1 to 4.7 m/s (2.1 m/s, on average). Ambient temperature ranged from 8°C to 14°C, whereas relative humidity was from 52% to 73%. Modeling the dispersion of pollutants for evaluating the urban air quality requires the meteorological data as well as the data on the source emission intensity (Szczygłowski and Mazur 2008). These parameters as well as traffic intensity data should be thoroughly investigated in further research.

In spring, when both mobile as well as fixed-site measurements were conducted, the mean hourly concentrations of PM_{10} and $PM_{2.5}$ – monitored by WIOS – equaled 13.1 and 26.8 µg/m^3 with standard deviations of 7.4 and 10.4 µg/m^3, respectively. It should be emphasized that these values were obtained via measurements conducted in a monitoring station located away from busy streets.

To sum up, the measurements performed in spring indicated that fluctuations of particle mass and number concentrations as well as changes in particle size distributions occurring in the monitored street in Lublin and its vicinity mainly depend on the intensity of vehicle traffic. These changes significantly affect the exposure to these air pollutants and may have negative health effects both on the commuters and pedestrians.

4.8.2 The summer measurement results

The measurements performed in summer enabled to track the changes in the particle number and mass concentrations along the route and in the established measurement points at different hours of the day. The results were diversified, depending on

Table 4.7 Exposure to particles by commuters and pedestrians in peak and off-peak traffic hours

	$PN_{0.1} \times 10^9$ pt/h	$PN \times 10^9$ pt/h	$PM_{2.5}$ μg/h	PM_{10} μg/h
Peak periods				
Commuters	4.2(2.2)3.6/1.3–9.4	4.8(2.4)4.2/1.6–10.2	4.7(1.6)4.4/2.1–7.9	29.6(10.7)28.4/11–67
Pedestrians	3.6(2.1)2.9/1.4–8.5	4.2(2.3)3.4/1.7–9.5	2.9(0.9)2.5/1.7–4.8	29.6 (8.6)27.4/13.8–48
Off-peak periods				
Commuters	1.4(0.4)1.3/0.7–2.8	1.7(0.5)1.6/1.0–3.2	2.5(0.5)2.4/1.6–4.6	12.6(3.2)11.8/7.9–21
Pedestrians	1.5(0.3)1.5/1.2–2.1	1.8(0.3)1.8/1.5–2.4	1.7(0.2)1.7/1.4–2.1	14.2(2.0)15.0/10.9–17

Arithmetic average(SD)median/min-max.

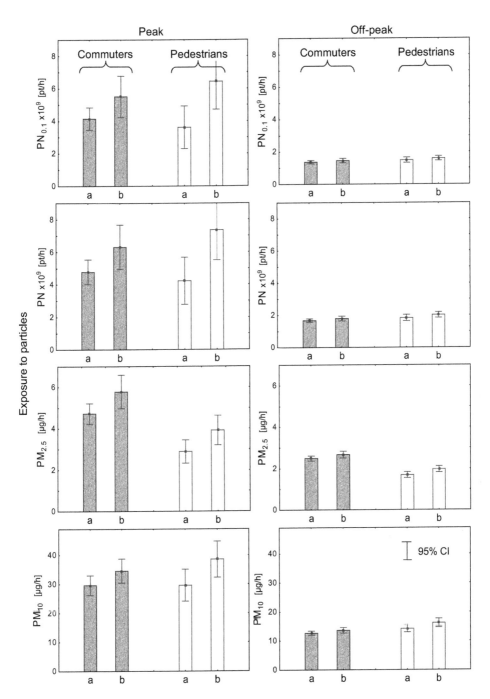

Figure 4.15 Mean doses of particles inhaled by pedestrians and commuters over a single hour spent in the investigated route during peak and off-peak times; a – entire route, b – part of the route with heaviest traffic; CI – confidence interval (data obtained by means of the Grimm instrument) (Połednik et al. 2018).

the instruments employed for measurements (Połednik et al. 2018). For greater consistency, only the results from peak (8:00) and off-peak (20:00) hours obtained using Grimm and P-Trak instruments are discussed further.

The variations of ultrafine and submicron particle number concentrations ($PN_{0.1}$ and PN_1) as well as mass concentrations of fine and coarse particles ($PM_{2.5}$, PM_{10}) in peak and off-peak hours are presented in Figure 4.16.

In summer, the average number concentration of $PN_{0.1}$ during the peak hours in the road section in the vicinity of 4-way TIs was approximately twice higher than in the off-peak hours. The average mass concentration of $PM_{2.5}$ was also roughly twice as higher in comparison to the off-peak hours. Comparable relationships were observed for other investigated aerosol particles and in terms of particle exposure.

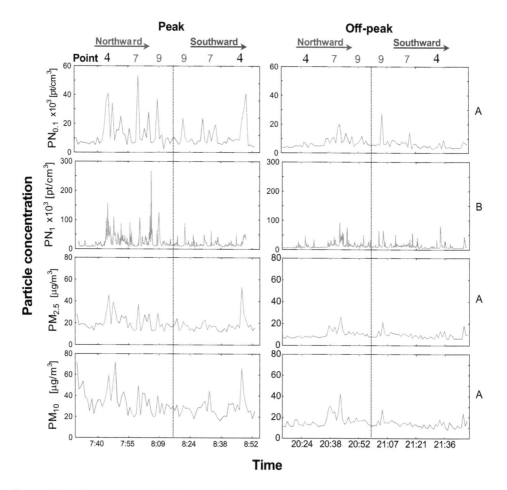

Figure 4.16 Time series of $PN_{0.1}$ and PN_1 concentrations as well as $PM_{2.5}$ and PM_{10} concentrations during peak and off-peak hours (4, 7 and 9 – measurement points near 4-way TIs, A – data obtained by means of the Grimm instrument, B – data obtained by means of the P-Trak instrument) (Piotrowicz and Połednik 2019).

Along the investigated route, the measured particle concentrations were characterized by a considerable variability. The most significant changes occurred at peak hours and in the vicinity of 4-way TIs (near points 4, 7 and 9). The greatest concentrations of PN_{01}, PN_1, $PM_{2.5}$ and PM_{10} at peak hours equaled 53.6×10^3 pt/cm^3, 266.8×10^3 pt/cm^3 and 53.0 and 72.1 μg/m^3, respectively. At off-peak hours, the highest concentrations were much lower and amounted to 27.2×10^3 pt/cm^3, 94.3×10^3 pt/cm^3 and 26.3 and 42.5 μg/m^3, respectively. At peak time, the average concentration of $PN_{0.1}$ in the section close to 4-way TIs reached $17.2 \pm 13 \times 10^3$ pt/cm^3 (mean ± standard deviation), i.e. approximately twice higher than at off-peak time. In the case of fixed-site measurement points, a similar relationship was noted in peak and off-peak hours. Figure 4.17 shows the average particle concentrations illustrating this relationship.

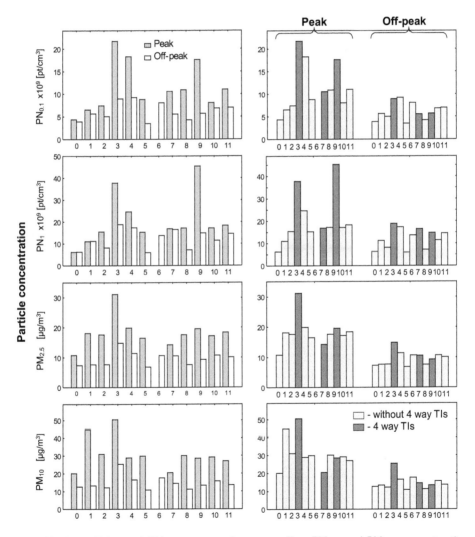

Figure 4.17 Mean $PN_{0.1}$ and PN_1 concentrations as well as $PM_{2.5}$ and PM_{10} concentrations during peak and off-peak hours in particular fixed-site measurement points (data obtained by means of the Grimm and P-Trak measurements).

The presented values of the measured particle concentrations are predominantly influenced by the intensity of traffic as well as the corresponding diversified particle emissions. Generally, in line with previous studies, the ultrafine and fine particles found in the air in the vicinity of communication routes mainly stems from the combustion of fuel in engines, whereas the coarse particles usually originate from non-exhaust sources, including road abrasion and brake and tire wear (Kumar and Goel 2016, Penkała et al. 2018).

Therefore, comparing the particle concentrations near 4-way TIs (i.e. parts with the highest traffic intensity) and other sections along the route seems to be highly relevant.

The statistical information pertaining to the particle number and mass concentrations from the measurements performed along the route as well as in fixed sites in direct adjacency to 4-way TIs and for the sections characterized by less intense traffic, both in peak and off-peak hours, is presented in Table 4.8.

Table 4.8 shows that the particle concentrations measured during the peak hours were higher. Moreover, the particle number concentrations in vicinity of 4-way TIs in peak and off-peak hours were, on average, approximately twice higher than in the part without the 4-way TIs. The particle mass concentrations were also about twice higher, on average, and a similar relationship was observed in the case of fixed-site measurements.

For commuters and pedestrians, the estimated exposure was based on the values obtained in the course of the on-road and fixed-site measurements. The estimated doses of particles inhaled by commuters and pedestrians following an hour on the investigated route, both in peak and off-peak hours, are presented in Table 4.9.

The details pertaining to the particle exposure as well as the estimated doses inhaled by commuters and pedestrians following an hour in the parts of the route with and without 4-way TIs, during the peak and off-peak periods, are presented in Figure 4.18.

The obtained findings prove that the exposure of commuters and pedestrian is dependent upon the location along the considered route as well as the intensity of traffic. The data indicate that greater doses or particles are inhaled by commuters and pedestrians during the peak hours. Moreover, greater particle number is inhaled in the vicinity of 4-way TIs, than in the parts with less intense traffic, both in peak and off-peak hours. As far as particle mass is concerned, the doses of PM_{10} (coarser particles) were greater in the part without 4-way TIs. During peak hours, the particle doses inhaled by commuters in the vicinity of 4-way TIs amounted to, on average, $4.3 \pm 3.3 \times 10^9$ pt/h ($PN_{0.1}$) or 2.9 ± 1.4 µg/h ($PM_{2.5}$). In the case of pedestrians, these values equaled $3.9 \pm 1.1 \times 10^9$ pt/h or 2.5 ± 0.4 µg/h. These doses are highly comparable to the ones estimated in the course of spring measurements (Połednik et al. 2018). In addition, they are similar to the results reported by Joodatnia et al. (2013). Slightly elevated doses inhaled by commuters are in line with the results reported by Kaur et al. (2005). However, commuters mostly drive in cars with closed windows and inhale filtered air.

4.8.3 The results of measurements conducted over four seasons

The results from the measurements performed over four seasons confirm that there is a clear relation between the traffic-related particle concentrations from the road and its proximity as well as the intensity of traffic. The concentrations of particles obtained from each sampling point in the course of individual runs were calculated

Table 4.8 Statistics of $PN_{0.1}$ and PN_I concentrations as well as $PM_{2.5}$ and PM_{10} concentrations for the on-route measurements in I – sections near 4-way TIs and II – sections with less intense traffic during the peak (8:00) and off-peak traffic hours (20:00) (Piotrowicz and Polednik 2019)

Part of the route	$PN_{0.1} \times 10^3$ pt/cm³	$PN_I \times 10^3$ pt/cm³	$PM_{2.5}$ μg/m³	PM_{10} μg/m³
Peak period				
I	9.7 (6.5)8.0/2.3–41.1	17.7 (11)14/7.8–106	18.4 (6.1)17.2/12–46	31.8 (10.3)30/18–71
II	17.2 (13)12.6/5.2–54	32.1 (30)21/8.6–267	21.8 (9.4)19.2/12–53	32.9 (14.1)29/17–72
Off-peak period				
I	5.6 (3.3)4.6/2.9–20	10.7 (7.1)8.2/5.3–65	8.6 (2.5)8.0/5.9–21.5	13.5 (3.2)13/8.1–28
II	8.3 (3.9)7.0/3.2–27	18.2 (14.3)13.2/6–94	12.1 (4.1)10.8/7.3–26	18.7 (7.1)17/11.7–43

Arithmetic average(SD)median/min-max.

Table 4.9 Exposure o² pedestrians and commuters in I – sections near 4-way TIs and II –sections with less intense traffic during the peak and off-peak hours (Piotrowicz and Polednik 2019)

Traffic period	$PN_{0.1} \times 10^9$ pt/h	$PN_I \times 10^9$ pt/h	$PM_{2.5}$ μg/h	PM_{10} μg/h
Commuters				
Peak I	2.6(1.9)2.0/0.6–10.3	4.9(4.0)3.5/1.7–39.5	2.7(0.9)2.5/1.7–6.5	28.9(9.9)27.5/16.1–64
II	4.3(3.3)3.2/1.4–13.5	8.1(7.6)5.6/2.3–67.2	2.9(1.4)2.4/1.7–7.5	27.1(11.2)22.3/14–59
Off-peak I	1.7(1.1)1.4/0.7–6.9	3.2(2.5)2.4/1.4–23.7	1.4(0.5)1.2/0.8–3.7	13.3(5.2)11.8/7.2–37.7
II	1.8(0.6)1.7/0.8–3.5	4.1(3.1)3.1/1.5–20.5	1.5(0.4)1.4/0.9–2.6	13.8(2.8)14.2/9.4–20.9
Pedestrians				
Peak I	2.5(1.3)2.1/1.1–5.5	4.4(2.3)4.1/1.6–9.5	2.6(0.8)2.5/1.5–4.4	29.1(8.8)26.6/17.8–44
II	3.9(1.1)4.4/2.7–4.6	7.3(3.7)6.2/4.2–11.4	2.5(0.4)2.8/2.0–2.8	23.0(4.2)25.3/18.2–26
Off-peak I	1.4(0.4)1.4/0.9–2.3	2.8(1.2)2.8/1.5–4.8	1.3(0.4)1.1/1.0–2.1	12.7(3.9)11.8/9.6–22.5
II	1.9(0.5)2.0/1.4–2.3	3.9(0.5)3.7/3.5–4.4	1.5(0.2)1.5/1.3–1.6	14.0(2.0)14.7/11.8–16

Arithmetic average(SD)median/min-max.

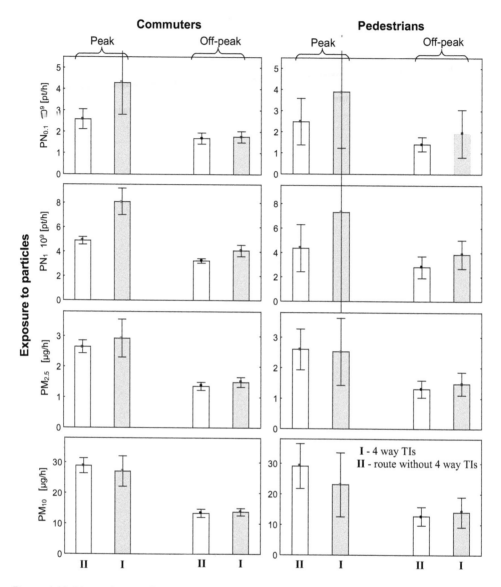

Figure 4.18 Mean doses of particles inhaled by pedestrians and commuters over a single hour spent in the investigated route during peak and off-peak hours; I – sections near 4-way TIs and II – sections with less intense traffic (based on the data from the Grimm and P-Trak measurements).

as differences of the measured total concentrations of size-fractionated particles and their background concentrations. The measurement of background concentrations was performed approximately 350 m away from the investigated route and within a certain distance from other streets, nearby a green area and a river, where the vehicle traffic was almost completely absent. Thus, the background levels were affected by traffic-related pollution to a limited degree or not at all.

At the beginning and after the end of each run, the average particle concentrations were measured over 5 minutes and then used as the background for each run. The average total number concentrations of ultrafine and submicron particles ($PN_{0.1}$, PN_1) and average total mass concentrations of fine and coarse particles ($PM_{2.5}$, PM_{10}) measured on the sidewalk in fixed-points during the morning rush hour (8 a.m.) and evening off-peak hour (8 p.m.) in four seasons are presented in Figure 4.19.

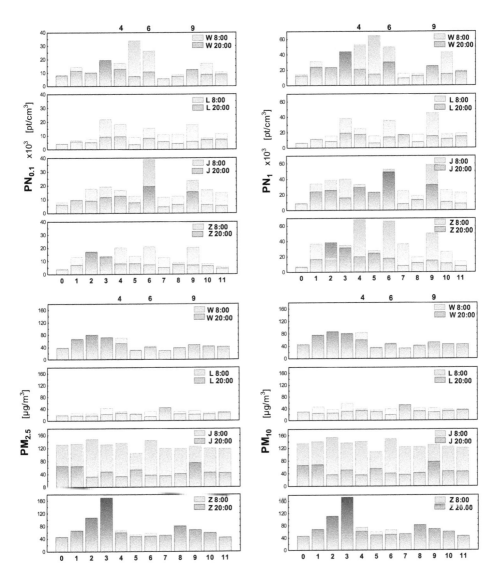

Figure 4.19 $PN_{0.1}$ and PN_1 concentrations as well as $PM_{2.5}$ and PM_{10} concentrations in particular fixed-site measurement points during peak (8 a.m.) and off-peak traffic hours (8 p.m.) in four seasons (W – spring, L – summer, J – fall, Z – winter) (based on the Grimm instrument data).

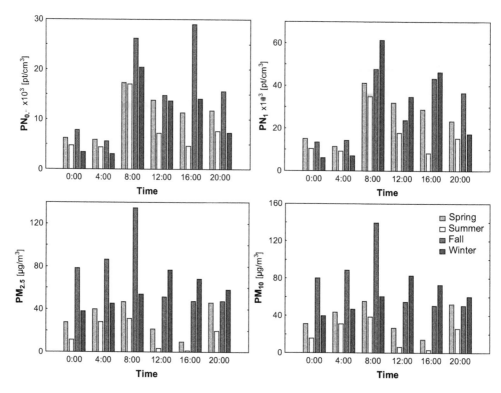

Figure 4.20 Mean $PN_{0.1}$ and PN_1 concentrations as well as $PM_{2.5}$ and PM_{10} concentrations for particular fixed-site measurement points near 4-way TIs at various times of the day and in different seasons.

Figures 4.20 and 4.21 show the average particle number concentrations ($PN_{0.1}$, PN_1) as well as particle mass concentrations ($PM_{2.5}$, PM_{10}) corresponding to the individual fixed-site measurement points near 4-way TIs (RS-I) and away from 4-way TIs characterized by less intense traffic (RS-II) at different times of the day and in different seasons. On the other hand, Figure 4.22 shows the particle number and mass concentrations corresponding to the entire route.

The presented graphs indicate that time of the day has a significant influence on the particle number and mass concentrations regardless of the season. The highest particle number and mass concentrations occur during the morning and afternoon rush hours. In the case of mass concentrations, the contribution of non-traffic sources is clearly visible. This is especially evident in fall, as well as in winter, when much higher mass concentrations of $PM_{2.5}$ and PM_{10} occur at night and in the morning (0 a.m., 4:00 and 8 a.m.). The emissions from household furnaces, which are the highest in these heating seasons, may be responsible for the increases in particle concentrations. The highest concentrations of particles at 8:00 may be connected with firing or adding fuel to household furnaces, which results in an increased emission of air pollutants from chimneys.

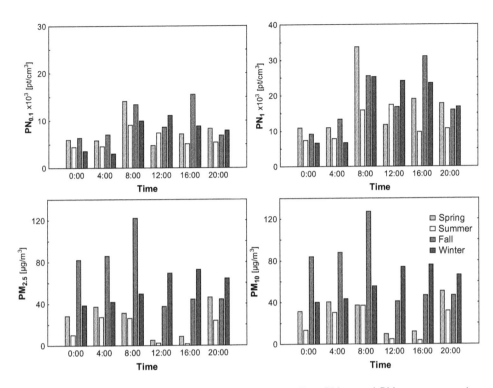

Figure 4.21 Mean $PN_{0.1}$ and PN_1 concentrations as well as $PM_{2.5}$ and PM_{10} concentrations for particular fixed-site measurement points away from 4-way TIs with less intense traffic at various times of the day and in different seasons.

Table 4.10 shows the average total number concentrations of $PN_{0.1}$, PN_1 as well as average total mass concentrations of ($PM_{2.5}$, PM_{10}) on the sidewalk in vicinity of 4-way TIs (RS-I) and away from them (RS-II) during the day as well as at night, in four seasons.

The table shows that the concentrations of particles on the sidewalk are highly influence by the intensity of traffic that is dependent on the time of the day as well as the season. It is possible that other sources of particles in the area may affect these concentrations, including domestic coal and biomass combustion appliances (Polednik 2013b, Marczak 2017). The concentrations of nearly all investigated particles were greater in RS-I and during the day than in RS-II and at night. The greatest total concentrations of $PN_{0.1}$, $PM_{2.5}$ and PM_{10} were observed in fall, during the day, on the sidewalk of 4-way TI (RS-I), reaching 23.3×10^3 pt/cm^3, 77.9 µg/m^3 and 81.5 µg/m^3, respectively. The greatest average total concentration of PN_1 was also found during the day, on the sidewalk of 4-way TI (RS-I), albeit in winter, reaching 47.4×10^3 pt/cm^3. In contrast, the lowest total concentrations were found on a sidewalk at night in winter, with comparable average levels in both sections: $PN_{0.1} - 4.6 \times 10^3$ pt/cm^3, $PN_1 - 10.1 \times 10^3$ pt/cm^3, $PM_{2.5} - 47.2$ µg/m^3 and $PM_{10} - 48.9$ µg/m^3.

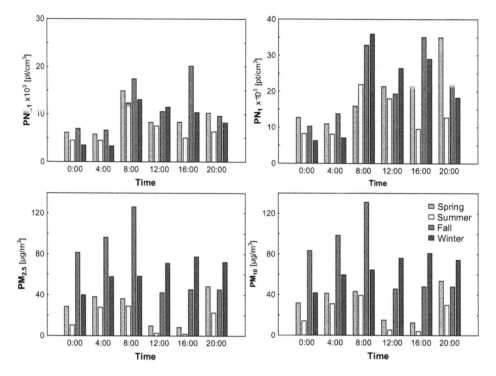

Figure 4.22 Mean PN$_{0.1}$ and PN$_1$ concentrations as well as PM$_{2.5}$ and PM$_{10}$ concentrations for all fixed-site measurement points along the route at various times of the day and in different seasons.

The shares of traffic-related particles in the total particle number and mass concentrations observed on the sidewalk of both RS-I and RS-II, at different times of the day and in different seasons, are shown in Figure 4.23.

The contribution of traffic-related particles to the total particle number and mass concentrations measured on the sidewalk in vicinity of intersections were mostly greater than away from them. In addition, they were higher during the day than at nighttime. Relatively high shares of the traffic-related particles to the total particle number concentrations were found in fall, winter and summer. The greatest concentrations of PN$_{0.1}$ and PN$_1$ concentrations were found in RS-I in winter at the morning rush hour (7:00–8:00), reaching 81.5%±0.3% and 88.1%±2.1%, respectively. As far as PM$_{2.5}$ and PM$_{10}$ mass concentrations are concerned, the greatest shares of traffic-related particles in RS-I were found in spring at the afternoon rush hour (15:00–16:00), amounting to 80.1%±9.1% and 59.7%±12.4%, respectively. Comparable relations involving elevated particle concentrations near TIs were found in numerous studies; for instance, increased concentrations of traffic-related particles were reported, for example, by Goel and Kumar (2015).

The study conducted by Ćwiklak et al. (2009) in Zabrze (Upper Silesia, Poland) in vicinity of an intersection with intensive traffic showed that the PM$_{2.5}$ and PM$_{10}$ concentrations were, on average, 46.9% and 44.9% higher than the urban background, respectively.

Table 4.10 Mean total $FN_{0.1}$ and PN_1 concentrations as well as mean total $PM_{2.5}$ and PM_{10} concentrations on the sidewalk of RS-I – section near 4-way TIs and RS-II – sections away from 4-way TIs with less intense traffic during the daytime (measurements at 8:00, 12:00, 16:00) and at night time (measurements at 20:00, 0:00, 4:00) for particular seasons

Season time	Route section	$PN_{0.1} \times 10^3$ pt/cm³	$PN_1 \times 10^3$ pt/cm³	$PM_{2.5}$ µg/m³	PM_{10} µg/m³
Spring					
Day	I	14.1(8.2)13.9/25.9	34.3(16.4)36.1/54.5	26.1(23.2)14.9/69.7	32.1(25.5)17.7/83.3
	II	9.8(7.9)7.3/33.6	21.8(15.0)19.5/63.3	16.0(13.2)10.4/38.8	21.0(14.4)16.5/47.5
Night	I	8.0(3.0)6.8/12.6	15.8(6.2)16.2/24.9	38.0(9.5)39.9/51.9	42.3(10.8)43.1/59.1
	II	6.7(2.0)6.0/11.3	13.2(5.9)13.1/27.0	37.3(13.1)33.2/79.2	40.9(14.4)36.1/85.1
Summer					
Day	I	10.0(6.1)7.2/18.2	18.6(13.5)15.4/45.3	13.1(15.2)3.4/32.4	17.2(17.9)6.4/40.3
	II	7.2(2.3)6.9/12.0	14.4(4.2)15.4/22.7	10.0(12.0)1.7/30.3	15.3(16.2)5.1/45.4
Night	I	5.6(1.8)4.7/9.2	11.6(4.3)12.8/17.4	20.0(8.0)20.5/29.3	24.2(7.8)26.9/32.7
	II	4.8(0.9)4.6/7.0	8.7(3.3)7.4/16.5	20.4(9.5)21.3/43.4	25.2(10.3)27.2/51.7
Fall					
Day	I	23.3(11.0)23.5/41	38.3(17.3)36.5/63.5	77.9(45.0)66.1/144	81.5(45.9)70.4/149
	II	12.5(5.6)12.3/30.4	24.5(12.3)22.3/65.2	68.0(40.9)44.4/146	71.7(41.9)46.2/153
Night	I	9.7(5.0)7.1/19.1	21.4(13.9)20.9/48.3	70.8(21.6)77.5/93.5	73.0(21.1)78.9/95.2
	II	6.7(1.6)6.3/11.1	12.8(8.1)8.3/30.4	70.7(20.6)82.5/89.9	72.9(20.5)84.9/91.5
Winter					
Day	I	16.1(5.8)19.7/23.1	47.4(17.5)50.0/68.0	66.4(13.8)65.5/82.7	72.1(14.3)73.4/89.7
	II	10.0(3.7)8.6/22.5	24.3(10.7)20.2/48.4	63.9(14.4)66.8/89.1	68.6(14.2)69.5/93.6
Night	I	4.6(2.1)3.8/8.0	10.2(5.8)7.8/19.6	47.2(11.1)41.5/67.5	48.9(11.3)43.5/68.9
	II	4.8(3.2)3.5/17.0	10.1(8.5)6.2/38.0	48.2(17.8)41.6/106	50.0(18.2)43.5/109

Arithmetic average(SD)median/maximum.

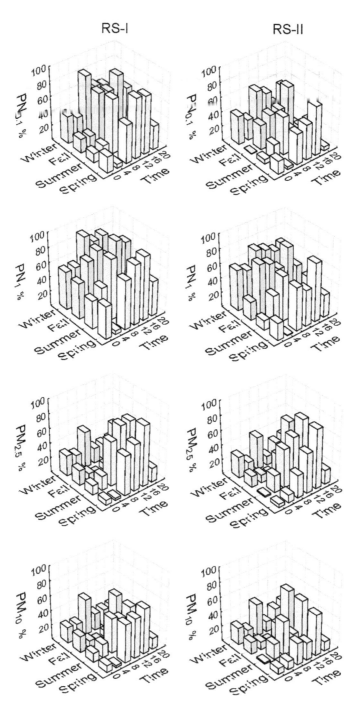

Figure 4.23 Share (%) of traffic-related particles to the total particle concentrations obtained on the sidewalk of RS-I and RS-II at various times of the day and in different seasons (Polednik and Piotrowicz 2020).

The smaller shares of traffic-related particles in total particle concentrations at night, apart from lesser traffic intensity, may stem from lower air temperatures increasing the relative humidity of air. As a consequence, condensation of water vapor may occur on generated and aggregated traffic-related particles, facilitating their deposition and limiting their spreading. Apart from that, water vapor condensation may also occur in the ground-level surfaces, which are usually cooler. This includes the road, sidewalk as well as the particles deposited on them, which hampers the resuspension of particles. The reduced mass concentrations of traffic-related fine and coarse particles could also result from higher humidity of the road surface due to greater frequency of precipitation in those seasons (Olszowski 2016, Zalakeviciute et al. 2018). The ratios of traffic-related particle concentrations on the sidewalk to the background concentrations were calculated for better characterization of the observations made. The ratios of traffic-related particle number concentrations $PN_{0.1}$, PN_1 as well as particle mass concentrations $PM_{2.5}$, PM_{10} observed on the sidewalk in vicinity of the 4-way TI characterized by most intense traffic during different times of the day and in different seasons, are shown in Figure 4.24.

Table 4.11 presents the statistical information related to the ratios of traffic-related particle number concentrations ($PN_{0.1}$, PN_1) as well as particle mass concentrations ($PM_{2.5}$, PM_{10}) observed on the sidewalk of RS-I and RS-II during the day and at night, in four seasons, to the background levels.

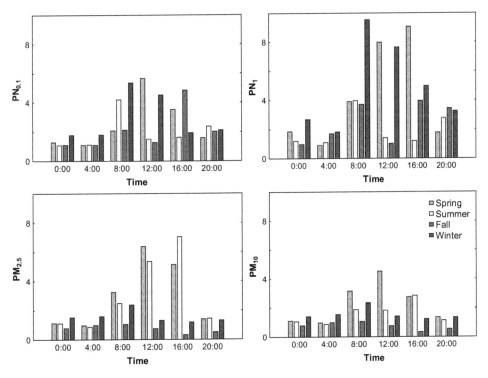

Figure 4.24 Ratios of traffic-related $PN_{0.1}$ and PN_1 concentrations as well as $PM_{2.5}$ and PM_{10} concentrations to background concentrations on the sidewalk in vicinity of four-way TI with the heaviest traffic at various times of the day and in different seasons.

Table 4.11 Ratios of traffic-related $PN_{0.1}$ and PN_1 concentrations as well as $PM_{2.5}$ and PM_{10} concentrations on the sidewalk of RS-I and RS-II during the day (at 8:00, 12:00, 16:00) and at night (at 20:00, 0:00, 4:00) in particular seasons to background levels

Season time	Route section	$PN_{0.1}$	PN_1	$PM_{2.5}$	PM_{10}
Spring					
Day	I	3.2 (3.2)2.2/8.5	4.2 (3.2)2.8/8.5	2.7 (2.8)2.2/8.1	1.3 (1.2)0.8/3.5
	II	1.2 (1.4)0.8/5.0	1.9 (2.0)1.6/7.2	1.3 (1.9)0.6/7.6	0.6 (0.5)0.5/2.4
Night	I	0.3 (0.2)0.3/0.6	0.6 (0.4)0.7/1.3	0.5 (0.7)0.3/2.2	0.2 (0.1)0.1/0.3
	II	0.2 (0.2)0.1/0.8	0.4 (0.4)0.4/1.5	0.2 (0.3)0.1/1.1	0.2 (0.2)0.1/0.9
Summer					
Day	I	1.3 (1.4)0.6/3.2	1.4 (1.4)0.2/6.3	2.7 (1.7)1.9/6.0	1.0 (1.1)0.7/4.0
	II	0.6 (0.5)0.4/1.5	0.7 (0.7)0.4/2.0	2.5 (3.8)1.1/13	0.9 (0.4)0.9/1.8
Night	I	0.4 (0.5)0.3/1.4	1.0 (0.7)1.2/1.8	0.2 (0.1)0.2/0.4	0.2 (0.1)0.1/0.4
	II	0.3 (0.2)0.2/0.8	0.5 (0.6)0.3/2.3	0.2 (0.3)0.2/1.5	0.2 (0.2)0.2/0.8
Fall					
Day	I	3.1 (2.1)3.6/7.5	3.8 (2.2)3.7/7.9	0.5 (0.5)0.3/1.4	0.5 (0.5)0.3/1.3
	II	1.2 (1.2)0.9/5.3	2.1 (1.7)1.8/8.1	0.5 (0.5)0.2/1.4	0.5 (0.5)0.2/1.3
Night	I	0.7 (0.7)0.5/2.1	2.3 (1.8)2.0/5.4	0.2 (0.2)0.2/0.5	0.2 (0.1)0.2/0.5
	II	0.3 (0.3)0.2/1.4	1.2 (1.7)0.5/6.7	0.2 (0.1)0.1/0.5	0.2 (0.1)0.1/0.5
Winter					
Day	I	2.5 (1.7)2.4/4.5	5.0 (2.6)4.7/8.5	0.5 (0.5)0.3/1.4	0.5 (0.4)0.4/1.4
	II	1.1 (0.9)1.0/3.4	2.0 (1.4)1.7/4.8	0.4 (0.4)0.3/1.7	0.4 (0.4)0.4/1.6
Night	I	0.7 (0.3)0.8/1.1	1.4 (0.7)1.5/2.2	0.3 (0.2)0.3/0.6	0.3 (0.2)0.3/0.5
	II	0.7 (0.8)0.5/3.5	1.4 (1.6)0.7/5.3	0.3 (0.4)0.3/1.3	0.3 (0.3)0.2/1.4

Arithmetic average(SD)median/maximum.

In most cases, the indicated ratios of particle concentrations, regardless of their size, were greater on the sidewalk of RS-I than on the sidewalk of RS-II. Moreover, they were much higher at daytime, than at night. The highest mean ratios of the traffic-related $PN_{0.1}$ to the background concentrations were observed on the sidewalk of RS-I in spring and fall, reaching 3.2 and 3.1, respectively. The highest ratios amounted to 8.5 and 7.5, whereas the coefficients of variation (CV) equaled 98% and 68%, respectively. The lowest ratios of the considered particle concentrations were found in winter, reaching 5.0. The mean ratio for spring was 4.2, whereas for fall it was 3.8. The smallest mean value, equaling 1.4, was observed in summer, as in the case of the $PN_{0.1}$ concentration ratios.

In terms of the traffic-related to background ratios of $PM_{2.5}$ and PM_{10}, the greatest mean values of mass concentrations during the day were found in spring and summer, reaching 2.7 and 2.7, respectively, for $PM_{2.5}$, whereas in the case of the PM_{10} concentrations, they amounted to 1.3 and 0.9, respectively. In spring and summer, the CV equaled 104% and 63%, respectively, for $PM_{2.5}$ and 92% and 110%, respectively, for PM_{10}. In turn, for fall and winter, the particles from traffic sources had relatively little influence on the concentrations of $PM_{2.5}$ and PM_{10} obtained on the sidewalk of both investigated sections. At daytime, they reached roughly one-third of the total measured $PM_{2.5}$ and PM_{10} concentrations, and they had an even lower value at night. In addition, no differences in the concentration ratios of the investigated particle were observed on the sidewalk of RS-I and RS-II, which may suggest that the particles which were not related to traffic had significant contributions in fall and winter. The seasonal variability of the mass concentration of urban air particles was studied by Rogula-Kozłowska et al. (2014) in three locations in Poland. They reported that in winter, the average concentration of $PM_{2.5}$ was almost twice as high as the mean during the non-heating period and three times as high as the mean found in summer. Similar findings were reported, for instance, by Marczak (2017) in Lublin, Poland, and by Meng et al. (2019) in Urumqi, China.

The contribution of traffic-related particles to the total doses of particles received by pedestrians on a sidewalk along a road with high traffic intensity in Lublin was estimated during the day and at night, in all four seasons. The greatest number and mass of traffic-related particles were inhaled during the day on the sidewalk in vicinity of 4-way TIs. In turn, the greatest contribution of traffic-related particles to the total number of received ultrafine particles was estimated in fall and reached 68.6%±8.5%. The greatest relative contribution of traffic-related $PM_{2.5}$ was observed in spring and reached 40.4%±2.7%. In turn, the lowest doses of traffic-related particles were inhaled by pedestrians in summer.

Table 4.12 presents the average daily contributions of the traffic-related particles to the total amount of particles which are inhaled by pedestrians on the RS-I and RS-II sidewalks in particular seasons.

In the presented research it was assumed that the distribution of pedestrian commute was 70% during the daytime (8:00 - 20:00) and 30% during the nighttime (20:00 - 8:00). The obtained results proved that the exposure of pedestrians is dependent upon the location along the route as well as the intensity of traffic which is related to it. The inhaled traffic-related particle number and mass concentrations were higher in vicinity of 4-way TIs, than away from them, independent of the season. In the case of RS-I, the mean daily shares of traffic-related $PN_{0.1}$ and PN_1 to the total

Table 4.12 Average shares (%) of traffic-related particles to the total amount of particles received by pedestrians on the sidewalk of RS-I and RS-II in particular seasons

Season	Route section	$PN_{0.1}$	PN_1	$PM_{2.5}$	PM_{10}
Spring					
	I	61.5±12.1	70.5±7.7	40.4±2.7	36.9±4.7
	II	40.0±9.2	53.0±13.1	28.0±7.4	22.8±6.8
Summer					
	I	39.9±25.4	41.8±31.7	35.7±14.1	29.2±9.9
	II	32.0±12.0	32.9±19.7	34.8±8.8	26.4±11.2
Fall					
	I	68.6±8.5	76.2±7.2	26.4±19.5	24.7±19.2
	II	44.6±13.6	62.7±10.2	20.9±19.8	20.1±18.4
Winter					
	I	64.7±11.7	79.9±5.9	28.2±12.0	29.2±10.9
	II	48.9±12.3	65.3±4.9	26.4±11.8	26.9±11.1

Mean ± standard deviation.

inhaled $PN_{0.1}$ and PN_1 from all sources were over 60% (in spring, fall and winter). The greatest average daily share of traffic-related $PN_{0.1}$ and PN_1 was observed in fall and winter, reaching 68.6%±8.5% and 79.9%±5.9%, respectively. In contrast, the lowest but still relatively substantial concentrations of $PN_{0.1}$ and PN_1 were found in summer, reaching 39.9%±25.4% and 41.8%±31.7%, respectively. As far as traffic-related particle mass concentrations are concerned, the highest concentrations of $PM_{1.5}$ and PM_{10}, reaching 40.4%±2.7% and 36.9%±4.7%, respectively, were inhaled by pedestrians in RS-I during spring. In turn, the lowest mass of these particles, i.e. 26.4%±19.5% and 24.7%±19.2%, respectively, were inhaled by pedestrians in RS-I in fall. The data presented show that much higher doses were inhaled by pedestrians in the parts of the route with greater traffic intensity. Moreover, a lower number of traffic-related particles was inhaled in the summer, whereas lower mass of traffic-related particles was received in the fall.

Figure 4.25 shows the fraction of traffic-related particles received by pedestrians on the sidewalk in particular seasons. In contrast, the estimated percentage contribution of traffic-related particles received by pedestrians on the sidewalk of RS-I and RS-II in particular seasons to the total amount of these particles received by them throughout the year is shown in Figure 4.26.

The greatest dose of traffic-related $PN_{0.1}$ received by pedestrians was observed in fall. The dose in RS-I amounts to 40.3%, whereas in RS-II, there was 34.3% of the yearly traffic-related dose of $PN_{0.1}$. The dose of traffic-related PN_1 is also substantial, reaching over 30% of the yearly dose. In the case of traffic-related particle mass, the highest amount is inhaled in winter, reaching approximately 35% and 37% of the average yearly dose of $PM_{2.5}$ and PM_{10}. In contrast, the lowest number and mass of traffic-related particles received by pedestrians occurred in summer, reaching less than 15%.

Direct comparison of the obtained results is challenging, because studies conducted by other authors in different urban areas in CEE mainly pertained to different health hazards caused by traffic. Juxtaposition with the studies performed in other areas of the world is not reliable due to disparate climate and meteorological conditions.

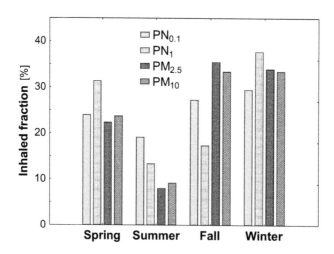

Figure 4.25 Share (%) of traffic-related particles received by pedestrians on the sidewalk in particular seasons.

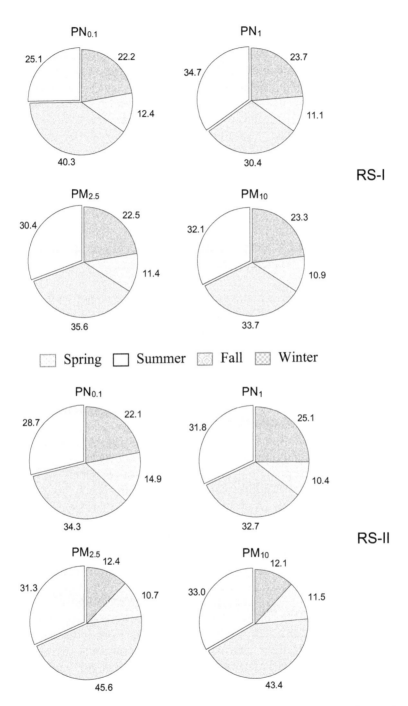

Figure 4.26 Share (%) of traffic-related particles in particular seasons to the total amount of particles received by them on the sidewalk of RS-I and RS-II throughout the entire year.

Nevertheless, the research conducted by Arkouli et al. (2010) in the urban area of Buenos Aires showed a clear seasonal variation in the average diurnal patterns of $PM_{2.5}$. The concentrations obtained in winter were greater compared to those is summer by a factor of 1.4, on average. Conversely, Alharbi et al. (2015) reported contrary results obtained from the study conducted in Riyadh, Saudi Arabia. The obtained PM concentrations were approx. 84% greater in summer, probably due to frequent dust storms at that time of year. Srimuruganandam and Shiva Nagendra (2010), who performed studies in vicinity of an urban road in Chennai, India, reported diurnal, weekly and seasonal cycles related to the concentrations of particular matter. Adeniran et al. (2017) showed seasonal variations and composition of suspended PM generated at large intra-urban TIs in Ilorin, Nigeria. During the dry season, the mean on-road respiratory deposition dose rates of PM_1, $PM_{2.5}$ and PM_{10} at TIs were approximately 24%, 9% and 25% higher, respectively, compared to the wet season.

4.9   MEASUREMENTS OF NON-TRAFFIC-RELATED AEROSOL PARTICLE CONCENTRATIONS

In Lublin, the possibility of conducting air pollution measurements with almost complete exclusion of traffic-related PM occurred in the course of the spring lockdown of 2020, which was introduced due to the spreading SARS-CoV-2 virus and COVID-19 pandemic. The MAPAL vehicle was used for performing mobile and fixed-site measurements along the route determined earlier. The measurement procedure and instruments were the same as in the previously conducted seasonal measurements of particle concentrations and estimation of exposure of drivers and pedestrians. Selected results of measurements were obtained on 16–17 April 2020 during 6 runs in 4-hour intervals (consecutive measurement runs at 8:00, 12:00, 16:00, 20:00, 0:00 and 4:00. The traffic intensity at daytime (i.e. runs at 8:00, 12:00 and 16:00) was relatively low and limited to several passing cars per minute. At night (20:00, 0:00 and 4:00), traffic was minimal, virtually non-existent. The pollutants in urban air almost completely originated from the sources located in low development – single and multi-family houses, which are equipped with furnaces mainly for the combustion of solid fuels (coal and timber).

Figures 4.27–4.30 present the changes in particle number concentrations PN_1, PN and shares of ultrafine particles $PN_{0.1}$, as well as changes in particle mass concentrations PM_1, $PM_{2.5}$ and PM_{10} in the course of particular runs.

Figures 4.31–4.33 present the comparison of particle number concentrations PN_1, PN and ultrafine particles $PN_{0.1}$, as well as particle mass concentrations $PM_{2.5}$ in the course of particular measurement runs in the spring of 2017 and 2020.

The data obtained in the course of particular runs were arranged depending on the year – spring of 2017 and 2020. Table 4.13 shows the basic statistical information on the PM and PN concentrations, measured using DustTrak and Grimm instruments. The average concentrations of PM_1 and $PN_{0.1}$ in the spring of 2017 reached 33.5 ± 22.1 $\mu g/m^3$ and $(8.2 \pm 5.8) \times 10^3/cm^3$, respectively (mean $\pm$ standard deviation). The values obtained in the spring of 2020 amounted to 25.5 ± 12.2 $\mu g/m^3$ and $(4.1 \pm 1.5) \times 10^3/cm^3$, respectively. In the case of the $PM_{2.5}$ and PN concentrations, the average values obtained in the spring of 2017 equaled 33.9 ± 23.3 $\mu g/m^3$ and $(9.8 \pm 6.4) \times 10^3/cm^3$, respectively. In the spring of 2020, they reached 26.1 ± 12.3 $\mu g/m^3$ and $(4.8 \pm 1.6) \times 10^3/cm^3$, respectively.

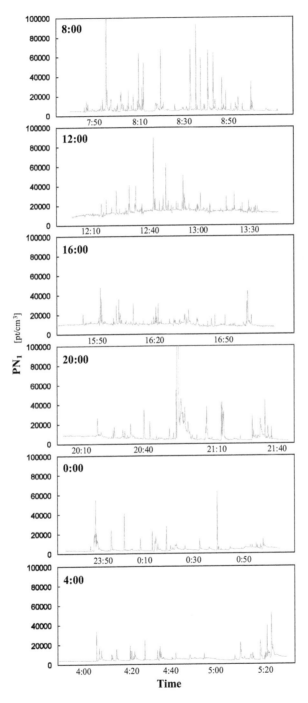

Figure 4.27 Fluctuations of PN_1 concentrations in the course of mobile and fixed-site monitoring in six consecutive measurement runs (0:00, 4:00, 8:00, 12:00, 16:00 and 20:00) (data obtained by means of the P-Trak instrument).

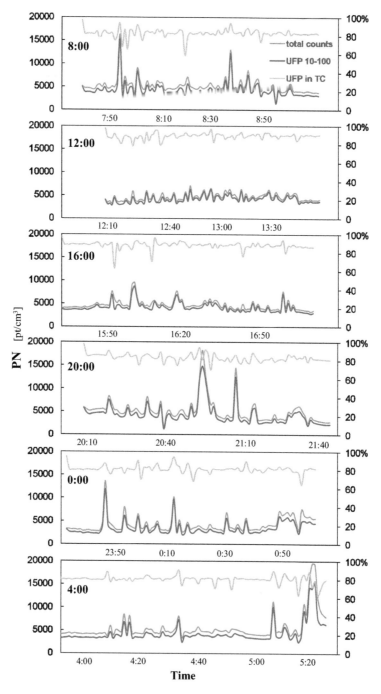

Figure 4.28 Fluctuations of PN concentrations in the course of mobile and fixed-site monitoring in 6 consecutive measurement runs (0:00, 4:00, 8:00, 12:00, 16:00 and 20:00) (data obtained by means of the Grimm instrument).

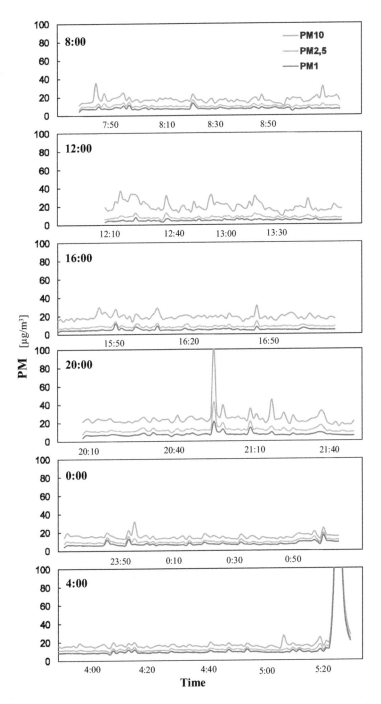

Figure 4.29 Fluctuations of PM$_1$, PM$_{2.5}$ and PM$_{10}$ concentrations in the course of mobile and fixed-site monitoring in 6 consecutive measurement runs (0:00, 4:00, 8:00, 12:00, 16:00 and 20:00) (data obtained by means of the Grimm instrument).

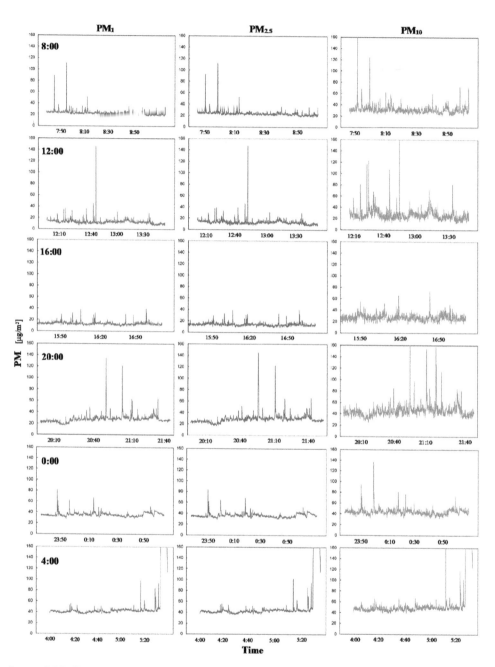

Figure 4.30 Fluctuations of PM₁, PM₂.₅ and PM₁₀ concentrations in the course of mobile and fixed-site monitoring in 6 consecutive measurement runs (0:00, 4:00, 8:00, 12:00, 16:00 and 20:00) (data obtained by means of the DustTrak instrument).

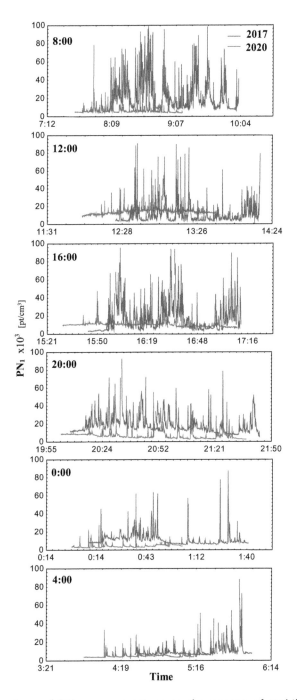

Figure 4.31 Comparison of PN₁ concentrations in the course of mobile and fixed-site monitoring in 6 consecutive measurement runs (0:00, 4:00, 8:00, 12:00, 16:00 and 20:00) in the spring of 2017 and 2020 (data obtained by means of the P-Trak instrument).

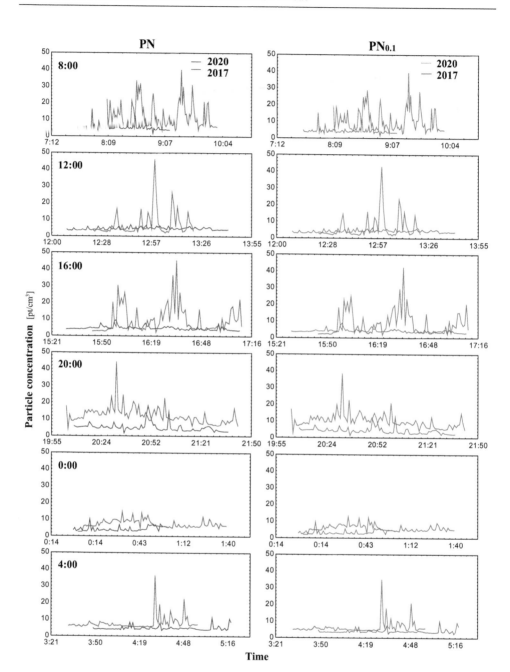

Figure 4.32 Comparison of PN and PN$_{0.1}$ concentrations in the course mobile and fixed-site monitoring in 6 consecutive measurement runs (0:00, 4:00, 8:00, 12:00, 16:00 and 20:00) in the spring of 2017 and 2020 (data obtained by means of the Grimm instrument).

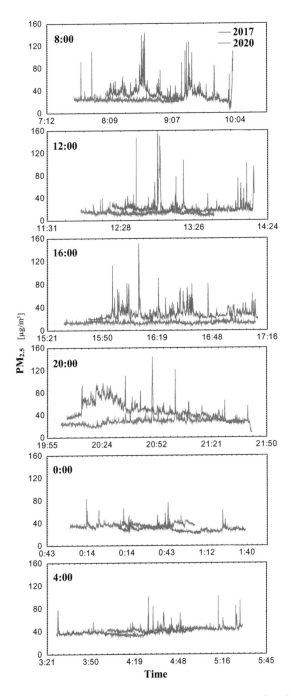

Figure 4.33 Comparison of PM$_{2.5}$ concentrations in the course of mobile and fixed-site monitoring in 6 consecutive measurement runs (0:00, 4:00, 8:00, 12:00, 16:00 and 20:00) in the spring of 2017 and 2020 (data obtained by means of the DustTrak instrument).

Table 4.13 Descriptive statistics for PM concentrations (in µg/m³) and PN concentrations (in ×10³/cm³) obtained in the course of the measurement runs in the spring of 2020 and 2017

Run	Particles	2020	2017
8:00	PM_1	23 (1.6) 22/27	37 (11.8) 35/83
	$PM_{2.5}$	23 (1.6) 23/28	38 (12.1) 35/84
	PM_{10}	31 (3.8) 31/46	46 (14.7) 43/105
	$PN_{0.1}$	44.2 (1.4) 3.9/12.2	10.8 (6.0) 9.1/28.8
	PN	55.0 (1.4) 4.8/13.0	12.9 (7.0) 10.6/33.4
12:00	PM_1	12 (2.4) 12/24	26 (34.3) 21/311
	$PM_{2.5}$	13 (2.6) 13/25	27 (37.8) 21/338
	PM_{10}	26 (5.4) 25/43	41 (96.4) 24/858
	$PN_{0.1}$	4.4 (0.8) 4.2/6.3	6.5 (72.2) 3.5/42.8
	PN	4.9 (0.8) 4.6/7.1	7.5 (7.8) 4.4/46.0
16:00	PM_1	14 (2.0) 13/23	29 (7.4) 28/66
	$PM_{2.5}$	14 (1.9) 14/24	29 (7.3) 29/67
	PM_{10}	28 (3.5) 27/37	34 (8.5) 33/72
	$PN_{0.1}$	4.2 (10.5) 4.0/8.7	8.6 (7.3) 5.6/42.3
	PN	4.8 (1.1) 4.6/9.5	10 (8.1) 6.8/45.4
20:00	PM_1	28 (5.8) 28/58	48 (15.5) 44/87
	$PM_{2.5}$	29 (6.1) 29/62	49 (15.7) 44/88
	PM_{10}	47 (12.4) 45/136	54 (17.2) 49/95
	$PN_{0.1}$	4.3 (2.2) 3.8/14.8	10.1 (4.5) 9.3/38.2
	PN	5.2 (2.6) 4.7/18.2	12.4 (5.1) 11.5/44.3
0:00	PM_1	36 (3.1) 36/44	30 (4.3) 30/45
	$PM_{2.5}$	37 (3.1) 36/44	31 (4.4) 30/45
	PM_{10}	44 (3.6) 43/53	34 (5.8) 34/56
	$PN_{0.1}$	3.5 (1.3) 3.0/9.6	6.3 (2.2) 6.2/12.8
	PN	4.4 (1.6) 3.8/10.2	7.7 (2.6) 7.9/14.6
4:00	PM_1	41 (1.9) 41/45	38 (5.4) 37/60
	$PM_{2.5}$	42 (2.0) 42/45	39 (5.7) 37/61
	PM_{10}	47 (2.5) 47/53	42 (7.5) 40/69
	$PN_{0.1}$	3.9 (0.8) 3.7/7.2	6.6 (4.4) 5.3/34.9
	PN	4.8 (0.9) 4.6/8.4	8.0 (4.5) 6.7/36.2

Arithmetic average (standard deviation) median/maximum.
Data obtained by means of the DustTrak and Grimm instruments.

In the case of PM_{10}, the average concentrations were 40.4 ± 50.9 and 37.4 ± 14.6 µg/m³ for the spring of 2017 and 2020, respectively. It should be emphasized that there were differences in the dates of measurements performed in 2017 and 2020, which could have a significant impact on the obtained results. In 2017, the measurements were performed on 4 and 5 April, whereas in 2020, they were done on 16 and 17 April. These periods differed in terms of the weather conditions and mean outdoor air temperature, which equaled 17.5°C and 9°C in 2017 (daytime/nighttime) as well as 16°C and 8°C in 2020. Moreover, in the first measurement period, the mean wind speed equaled 9.5 km/h, as opposed to 24.1 km/h in 2020, which could have affected the intensity of emissions from non-traffic-related sources as well as the amount of particles resuspended from the road and its vicinity. This could have contributed to higher overall average concentration of particles obtained from test runs in 2020. It also needs to be emphasized that in the course of the 4:00 night run in the spring of 2020, marked increases in the particle number and mass concentrations were noted towards the finish. These increases were accounted for in the calculations and resulted from temporary intensive emissions from household furnaces fired in the morning.

The comparison of aerosol particle concentrations in the air along the monitored route in the spring of 2017 (with normal vehicle traffic) and 2020 (with the traffic limited by COVID-19 pandemic) illustrates the impact of traffic-related pollution on the air quality in Lublin. Although the weather conditions during the test in 2020 were conducive to higher particle concentrations, the obtained values were clearly lower than in 2017. Moreover, some runs (8:00 and 20:00) yielded lower particle mass concentrations. Once again, it should be noted that intensive emissions occurred towards the end of the 2020 run at 4:00, which had a significant impact on the reported particle mass concentrations.

The concentration of certain chemical pollutants in the urban air measured by the Chief Inspectorate for Environmental Protection and the concentrations of aerosol particles obtained by the automatic measurement station in Lublin throughout the considered test period in 2017 and 2020 are presented in Figure 4.34.

During the period of mobile and fixed-site measurements conducted in the course of the study, the average hourly concentrations of PM_{10} and $PM_{2.5}$ reached 13.1 and 26.8 $\mu g/m^3$, with standard deviations of 7.4 and 10.4 $\mu g/m^3$, respectively. It should be noted that these results pertain to the measurements conducted in a monitoring station that is not located in direct vicinity of busy streets.

Taking into account the unprecedented reduction of urban traffic as a result of the measures introduced during the COVID-19 pandemic, it was predicted that a substantial reduction in the levels of air pollutants would occur in Lublin. The comparison of the presented results indicates that the concentrations of certain air pollutants measured in the air quality monitoring station in 2020 are clearly lower than their concentrations measured in 2017. This pertains mainly to such pollutants as CO, C_6H_6 and NO_2. However, it cannot be confirmed in the case of the remaining pollutants, including SO_2, O_3, $PM_{2.5}$ and PM_{10}. This is probably connected with different meteorological conditions on the considered test days, which may significantly affect the emissions of pollutants to the urban air, as well as their removal.

Similar results of reduced concentrations of some air pollutants during the unprecedented decrease in the urban traffic volume resulting from the measures introduced during the COVID-19 pandemic were obtained in other cities worldwide. For instance, the research carried out in China indicated that the ambient concentrations of $PM_{2.5}$, NO_2, SO_2 and CO during the pandemic dropped by approximately 30%–40%, 30%–60%, 20%–30% and 30%, respectively, whereas O_3 increased by roughly 10% in comparison with the same period in previous years or the preresponse period within the same year (Bauwens et al. 2020, Li et al. 2020, Shi and Brasseur 2020, Xu et al. 2020). The satellite data pertaining to NO_2 indicated that 20%–38% decreases of NO_2 occurred in 2020 in the cities of Western Europe and major Northeastern United States, compared to 2019. However, Iran, a country which was significantly affected by the COVID-19 pandemic, did not indicate clearly lower concentrations of NO_2 (Bauwens et al. 2020). In the United States, the preliminary findings of a study involving the data on daily $PM_{2.5}$ and O_3 concentrations acquired from US EPA (Environmental Protection Agency) show that the levels of $PM_{2.5}$ were approximately 10% greater than anticipated, whereas the levels of O_3 were roughly 7% lower than

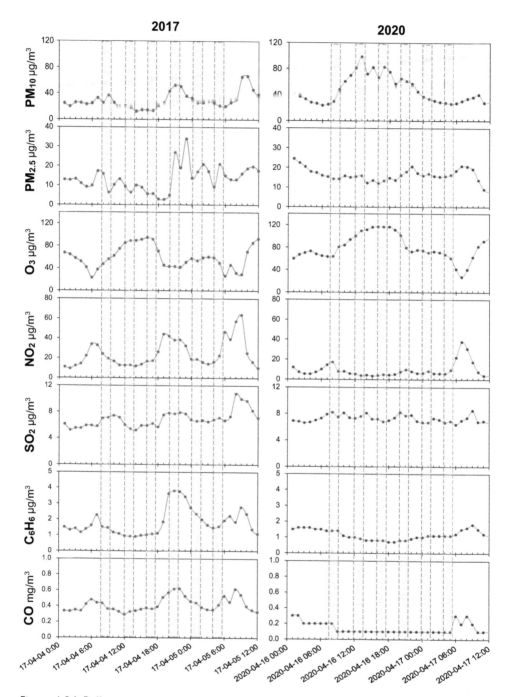

Figure 4.34 Pollutants concentrations obtained in air quality monitoring station in Lublin during the test period in 2017 and 2020.

anticipated following the post-"Stay Home Order" period, in comparison to the same period in previous years (Bekbulat et al. 2020). Moreover, this study indicates that NO_2 in Seattle reached levels that were 20% lower than anticipated. The significant diversification of the findings is attributed to the role of factors not related to the COVID-19 pandemic (for instance weather and emissions from regional non-traffic-related events) and may disturb these short-term trends (Arya 1999, Ault et al. 2009, Chen et al. 2017, Husar and Renard 1998, Zhao et al. 2020). The influence of weather as well as regional events can be partially compensated through comparison of the current data with the data from previous years (Bekbulat et al. 2020). Zhao et al. (2020) attempted to exclude the weather-related impact on the change in air pollution by applying the Community Multiscale Air Quality (CMAQ) model and Weather Research and Forecasting (WRF) model, which were combined with certain attributing analyses. Unfortunately, as it was mentioned in the study, WRF and other regional models encounter problems while simulating the evolution of the boundary layer (Banks et al. 2015, Hu et al. 2010) that plays a major role in formation and dispersion of air pollution. Hence, the influence of weather cannot be accurately determined using the WRF model. Thus, developing an approach to fully account for the aforementioned factors is essential.

4.9.1   Exposure to traffic-related and non-traffic-related PM

Exposure to traffic-related air pollution in the form of PM was evaluated using the contamination factor (CF) values calculated from the following equation:

$$CF = \frac{(C - \overline{C_b})}{\overline{C_b}} \times 100 \tag{4.2}$$

where C is the particle concentration in the ambient air obtained during mobile measurement and $\overline{C_b}$ is the average background particle concentration.

The inhaled dose of traffic-related particles (ID) by road users (drivers and pedestrians) was estimated using a modified equation presented in Polednik and Piotrowicz (2020):

$$ID = V_T \times f \times (C - \overline{C_b}) \times t \tag{4.3}$$

where V_T is the tidal volume, f is the breathing rate, and t is the time of exposure. It was assumed that for an average road user (for male adult) under light exercise conditions, V_T amounts to $800\,cm^3$ per breath and f is equal to 21 per minute (Hinds 1999).

Total deposited dose (DD) of particles in the respiratory systems of road users was calculated from the formula:

$$DD = V_T \times f \times DF \times C \times t \tag{4.4}$$

and deposited dose of traffic-related particles (DD_T) was obtained from the formula:

$$DD_T = V_T \times f \times DF \times (C - \overline{C_b}) \times t \tag{4.5}$$

where DF is the total deposition fraction estimated based on the simplified equations fitted to the ICRP model (ICRP 1994a, 1994b) describing regional depositions, namely in the head airways and tracheobronchial and alveolar region of the lungs. DF is the sum of these regional depositions (Hinds 1999) and can be given by:

$$DF = IF \times (A + B + C) \tag{4.6}$$

where IF constitutes the inhalable fraction which is computed as:

$$IF = 1 - 0.5 \left(1 - \frac{1}{1 + 0.00076 \, d_p^{2.8}} \right) \tag{4.7}$$

$$A = 0.058$$

$$B = \frac{0.911}{1 + \exp(4.77 + 1.485 \ln d_p)} \tag{4.8}$$

$$C = \frac{0.943}{1 + \exp(0.508 - 2.58 \ln d_p)} \tag{4.9}$$

and d_p is median diameter of particles.

Some results of exposure to PM by road users is presented in the following part.

Figure 4.35 shows total and traffic-related number concentrations of ultrafine particles $PN_{0.1}$ and total and traffic-related mass concentrations of submicron particles PM_1 measured in daytime and nighttime in 2017 and 2020.

It is clearly visible that during the lockdown period in 2020, lower concentrations of $PN_{0.1}$ and PM_1 occurred in ambient air, both during the day and at night.

The average percentage contamination of ambient air by traffic-related particles as well as their average inhaled doses by drivers and pedestrians during one hour at the considered times of day in 2017 and 2020 are presented in Table 4.14.

The table shows that regardless of the time of the day, the contamination factor values and inhaled doses of traffic-related particles were multiple times lower in 2020, compared to 2017. The mean contamination with ultrafine particles $PN_{0.1}$ and mean inhaled number of these particles were in the daytime (based on the data from the runs performed at 8:00, 12:00 and 16:00) 5.7(1.6)- and 6.1(0.5)-fold lower in 2020, respectively; the greatest reductions in the value of both parameters occurred in the afternoon peak hours and equaled 6.7. In the evening and at night, the mean air contamination by ultrafine particles $PN_{0.1}$ and their inhaled doses were 1.6(1.0)- and 2.4(0.5)-fold lower in 2020, respectively, than in 2017.

In the case of the traffic-related submicrometer particles PM_1, the mean air contamination by these particles was in daytime 3.8(1.8)-fold lower, and the mean inhaled mass of these particles was 4.9(1.4)-fold lower in 2020 than in 2017. The greatest reduction of the contamination values, amounting to 5.4, occurred in the morning peak hours (8:00), whereas the highest reduction of the inhaled doses, reaching 5.9, was noted in the afternoon peak hours (16:00). In the evening and at night (based on

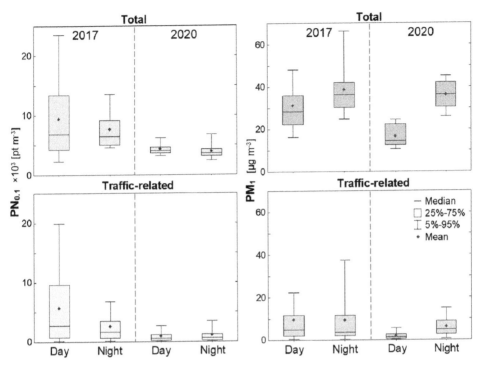

Figure 4.35 Total and traffic-related particle number concentrations $PN_{0.1}$ and particle mass concentrations PM_1 in day and night in 2017 and 2020.

the data from the runs at 20:00, 0:00 and 4:00), the mean contamination values were 1.3(0.4)-fold lower, and the mean inhaled mass of these particles was 1.3(0.5)-fold lower than in 2020.

Both in 2020 and 2017, the highest values of contamination factors determined for particle mass concentrations (PM) and estimation of their inhaled doses occurred in the evening (20:00). It can be purported that this is connected with the relatively highest reduction of background particle mass concentrations. Generally, at that time of the day, the emissions from domestic furnaces are less intense. In these furnaces, following the cooking period in the afternoon, the heat is usually maintained or promptly damped. This condition lasts until addition of fuel or another firing before sleep.

In order to estimate the doses of particles deposited in lungs of road users per unit time, the measurement data pertaining to particle size distributions were employed. Figure 4.36 presents number-size distributions of total and traffic-related particles $PN_{0.1}$ and mass-size distributions of total and traffic-related particles, with PM_1 representative for the considered measurement periods – days and nights in 2017 and 2020.

Table 4.15 presents the values of median diameters, deposition fractions, average concentrations of total and traffic-related particle fractions $PN_{0.1}$ and PM_1 estimated for the considered measurement periods – days and nights in 2017 and 2020. The average doses of these particles deposited in the lungs of road users at that time are shown as well.

The table shows that, generally, doses of particles deposited in the lungs of road users during the COVID-19 lockdown were markedly lower, both during the day and at night, than in 2017. In either case, they were lower at night. The greatest reductions in the deposited doses occurred during the day. Over an hour, the total number of deposited particles $PN_{0.1}$ and the total mass of deposited particles PM_1 were 2.2- and 2.1-fold lower, respectively. In turn, the number and mass of deposited traffic related particles were over 5-fold lower and the ratio for $PN_{0.1}$ amounted to 5.3, whereas for PM_1 it was 5.2. At night, the reductions of the deposited considered particle fractions were slightly lower (ratio ~2); the lowest reductions (ratio ~1.3) occurred in the case of total and traffic-related mass of particles PM_1.

Doses of deposited total and traffic-related particle fractions $PN_{0.1}$ and PM_1 over a single hour during the day and at night in 2017 and 2020 are presented in Figure 4.37.

Table 4.14 Values of the traffic-related PM contamination factor (in %) as well as doses inhaled by road users ID (in $\times 10^9$ pt/h for PN and in μg/h for PM) at different times of the day in 2017 and 2020

Time	Particle fraction	2017		2020	
		Contamination	Inhaled dose	Contamination	Inhaled dose
8:00	$PN_{0.1}$	125(135) 76/708	6.1(6.6) 3.7/34.5	33(49) 19/341	1.1(1.7) 0.6/11.9
	PN	20(123) 76/581	7.1(7.3) 4.5/34.3	31(41) 19/288	1.3(1.7) 0.8/12.2
	PM_1	38(40) 28/209	10.1(10.8) 7.6/56.4	7(6) 6/28	1.8(2.6) 1.4/18.7
	$PM_{2.5}$	38(40) 28/210	10.3(10.9) 7.7/57.5	7(6) 6/30	1.8(2.6) 1.3/18.9
	PM_{10}	34(37) 22/207	11.6(12.7) 7.7/71.3	9(7) 8/29	3.7(4.5) 2.8/36.5
12:00	$PN_{0.1}$	163(257) 40/1353	4.8(7.6) 1.2/40.2	25(21) 19/83	0.9(0.7) 0.7/2.9
	PN	146(226) 38/1176	5.3(8.2) 1.4/42.8	25(20) 19/85	1.0(0.8) 0.7/3.3
	PM_1	54(213) 11/1633	9.7(38.5) 1.9/296	28(23) 26/138	2.9(2.3) 2.7/14.2
	$PM_{2.5}$	57(228) 11/1758	10.4(41.8) 2.0/322	27(22) 25/134	3.0(2.4) 2.7/14.4
	PM_{10}	106(491) 13/3834	23.2(108) 2.9/843	32(28) 23/117	6.9(6.1) 5.0/25.4
16:00	$PN_{0.1}$	173(215) 73/1176	5.8(7.2) 2.5/39.3	26(28) 17/150	0.9(1.0) 0.6/5.3
	PN	166(203) 80/1056	6.6(8.0) 3.2/41.8	26(28) 17/143	1.0(1.1) 0.7/5.6
	PM_1	47(36) 43/230	9.5(7.3) 8.7/46.5	12(12) 9/82	1.6(1.5) 1.2/10.6
	$PM_{2.5}$	49(36) 44/232	9.8(7.4) 9.0/47.0	12(12) 10/80	1.6(1.5) 1.3/10.8
	PM_{10}	62(38) 57/240	13.3(8.1) 12.3/51.4	13(11) 10/44	3.4(2.8) 2.7/11.4
20:00	$PN_{0.1}$	38(50) 27/377	3.1(4.0) 2.1/30.5	57(82) 30/412	1.6(2.4) 0.9/12.0
	PN	38(47) 27/359	3.7(4.5) 2.6/34.9	54(77) 29/423	1.9(2.7) 1.1/14.9
	PM_1	70(53) 57/202	20.2(15.5) 16.5/58.7	57(28) 51/207	10.9(5.4) 9.8/39.6
	$PM_{2.5}$	70(53) 57/200	20.3(15.5) 16.6/58.8	56(29) 50/213	11.1(5.7) 9/9/42.5
	PM_{10}	60(48) 45/174	20.9(16.8) 15.6/60.6	45(39) 38/309	15.1(13) 12.6/103
0:00	$PN_{0.1}$	84(54) 67/278	2.9(1.8) 2.3/9.5	31(32) 21/197	1.0(1.1) 0.7/6.4
	PN	81(51) 64/250	3.4(2.1) 2.7/10.5	29(29) 20/155	1.2(1.2) 0.8/6.2
	PM_1	12(9) 12/58	3.4(2.5) 3.4/16.6	7(4) 7/18	2.8(1.6) 2.7/6.8
	$PM_{2.5}$	12(9) 12/58	3.5(2.6) 3.4/16.9	7(4) 7/18	2.8(1.6) 2.7/6.5
	PM_{10}	14(11) 12/79	4.2(3.6) 3.8/24.9	7(4) 6/19	3.0(2.0) 2.7/8.5
4:00	$PN_{0.1}$	34(80) 10/571	1.8(4.2) 0.5/30.0	23(32) 16/207	0.8(1.1) 0.5/6.8
	PN	31(67) 10/466	2.0(4.3) 0.7/30.0	20(27) 13/169	0.8(1.1) 0.6/7.1
	PM_1	10(10) 8/67	3.5(3.5) 2.9/24.3	10(8) 9/45	4.1(3.1) 3.6/17.6
	$PM_{2.5}$	10(10) 8/68	3.6(3.7) 3.0/24.8	10(8) 9/47	4.1(3.2) 3.6/18.6
	PM_{10}	11(12) 8/74	4.6(4.9) 3.1/29.6	12(11) 10/68	5.4(5.0) 4.6/30.4

Arithmetic average(standard deviation) median/maximum.
On the basis of the data obtained from DustTrak and Grimm.

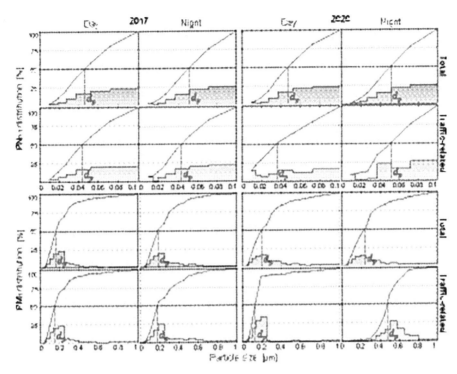

Figure 4.36 Number-size distributions of total and traffic-related particles $PN_{0.1}$ and mass-size distributions of total and traffic-related particles PM_1 in considered days and nights in 2017 and 2020.

Table 4.15 Median diameters d_p, deposition fractions *DF*, average concentrations of total *C* and traffic-related $C - \bar{C}_b$ particles and deposited doses of total *DD* and traffic-related DD_T particle fractions $PN_{0.1}$ and PM_1 in 2017 and 2020

| Time | Particle fraction | 2017 | | | | | | | |
| | | Total | | | | Traffic-related | | | |
		d_p μm	DF	C $\mu g/m^3$ $(\times 10^3$ pt/ $cm^3)$	DD $\mu g/h$ $(\times 10^9$ pt/h)	d_p μm	DF	C $\mu g/m^3$ $(\times 10^3$ pt/ $cm^3)$	DD $\mu g/h$ $(\times 10^9$ pt/h)
Day	$PN_{0.1}$	0.046	0.47	9.39	4.44	0.044	0.48	5.69	2.78
	PM_1	0.165	0.16	31.27	5.17	0.131	0.20	9.74	1.93
Night	$PN_{0.1}$	0.052	0.43	7.71	3.34	0.043	0.49	2.60	1.29
	PM_1	0.188	0.15	38.84	5.88	0.178	0.16	9.31	1.46
		2020							
Day	$PN_{0.1}$	0.047	0.46	4.34	2.02	0.036	0.55	0.96	0.53
	PM_1	0.202	0.14	16.77	2.43	0.151	0.18	2.08	0.37
Night	$PN_{0.1}$	0.051	0.44	4.02	1.76	0.051	0.44	1.14	0.50
	PM_1	0.242	0.13	36.32	4.85	0.473	0.16	6.21	0.98

C and $C - \bar{C}_b$ in $(\times 10^3$ pt/cm$^3)$, *DD* and DD_T in $(\times 10^9$ pt/h) for $PN_{0.1}$.

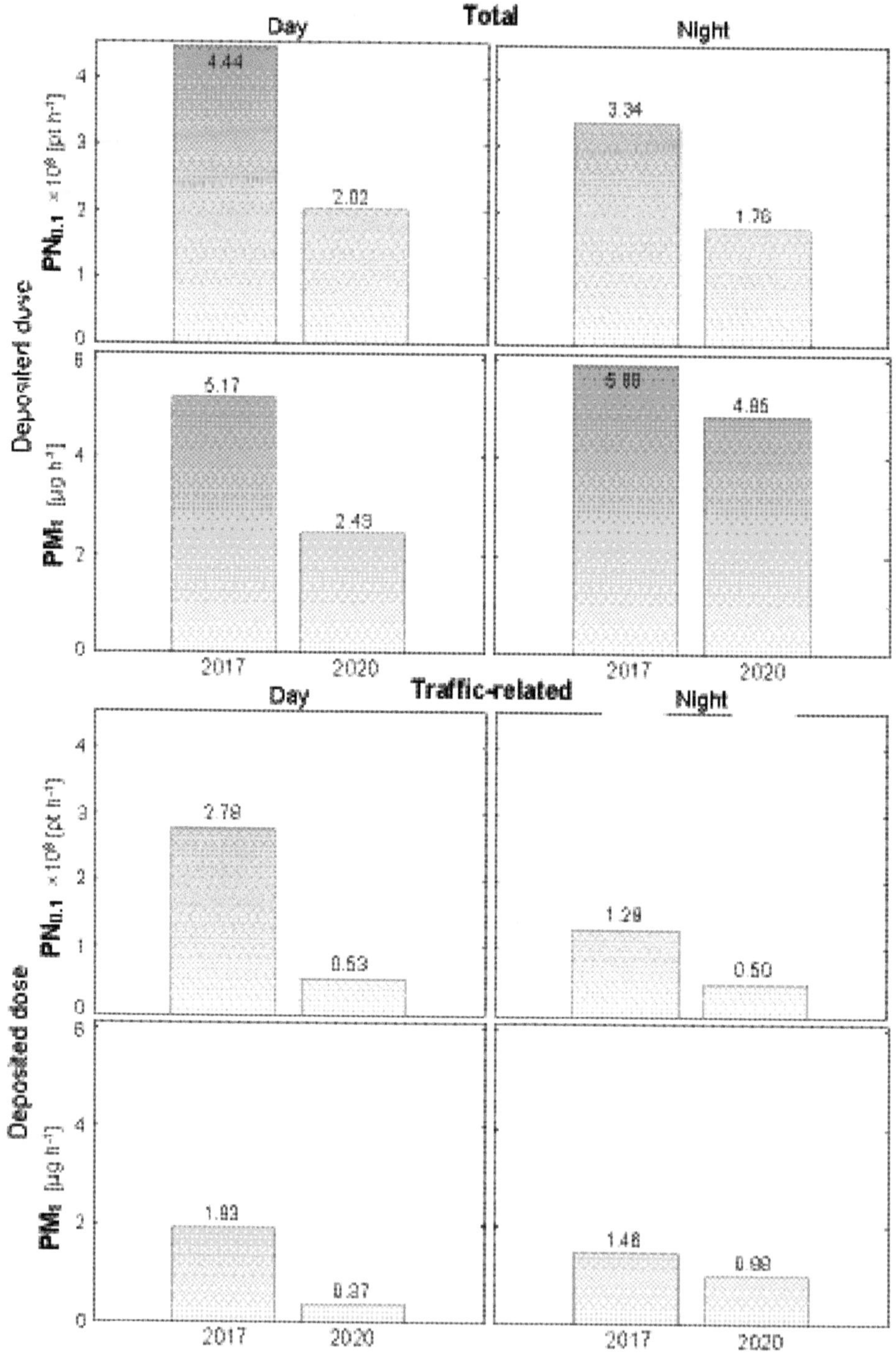

Figure 4.37 Deposited doses of total and traffic-related particle fractions $PN_{0.1}$ and PM_1 in the lungs of road users in the given day and night in 2017 and 2020.

The presented data indicate that the lockdown period contributed to a significant improvement of the ambient air quality in Lublin. This occurred as a result of a greatly reduced volume of vehicular traffic and, thus, decreased amount of the emitted traffic-related pollution. Undoubtedly, the beneficial health effects for road users and Lublin residents in general would be greater if other intensive sources of urban air pollution were eliminated or mitigated. The emissions connected with residential activity, predominantly the emissions from domestic furnaces, are especially problematic. During the heating season (autumn, winter and early spring), these emissions are the main factor deteriorating the quality of air in Lublin.

4.10 MEASUREMENTS OF TRAFFIC-RELATED POLLUTION IN OTHER CITIES AND REGIONS OF POLAND

The studies on traffic-related pollution were conducted, for instance, in Kraków, in the vicinity of main roads. They indicated that the vehicular traffic significantly contributes to the increased concentration of aerosol particles in their proximity (Wróbel et al. 2000). At the shortest distances from the road (up to 5 m), the share of coarse particles with the size of 1.9–72 µm in the total number of measured particles amounted to 53%–68%. At distances up to 150 m from the considered roads, the vehicular traffic was responsible for 80% of coarse particles present in the air. As the distance increased from 150 to 200 m, the concentration of these particles decreased almost in half. At a distance of 1,500 m, the share of traffic-related particles in the air reached only 20%. It was also noted that the lifespan of these particles is long enough to significantly affect the air quality in a relatively large urban area.

The comparison studies of particle concentrations in the vicinity of a busy intersection related to the urban background, measured away from busy roads, were also conducted in Zabrze (Rogula-Kozłowska et al. 2008). The concentrations of PM_{10} and $PM_{2.5}$ by the intersection were 17.7 and 10.2 $\mu g/m^3$ higher, respectively.

The unfavorable impact of street canyons on the concentration of particles was confirmed in the course of studies conducted beside a busy urban road in Gliwice (Grynkiewicz Bylina et al. 2005). The mean daily concentration of PM_{10} in the street canyon was 40 $\mu g/m^3$ higher than at the distance of 100 m.

According to the studies conducted in Kraków, the changes in weather conditions, including the windspeed and the height of the mixing layer, constitute a significant factor affecting the concentrations of traffic-related pollution and the quality of urban air (Oleniacz et al. 2016). The meteorological data and information on the intensity of emission sources are also important in modeling the propagation of traffic-related pollution and assessing the quality of urban air (Szczygłowski and Mazur 2008, Holnicki et al. 2017).

Other studies on particle concentration changes recorded in measurement points located along main roads conducted in Poland reported that the concentration of traffic-related particles was halved at a distance of 150–200 m from the roads (Wróbel et al. 2000). Skubacz (2009) reported that substantial numbers of submicron particles are generated in the vicinity of busy roads. Rogula-Kozłowska et al. (2008) stated that the vicinity of busy intersections also significantly affects the particle concentration levels. Elevated concentrations of particles also occur in street canyons (Grynkiewicz Bylina et al. 2005).

Kozielska et al. (2013) studied the ambient concentration of particle-bound polycyclic aromatic hydrocarbons (PAHs) in Zabrze and Ruda Śląska, two industrial cities in Upper Silesia. The 24 hours $PM_{2.5}$ were collected over 24 hours, both at busy roads (or crossroads) and an urban background site. Moreover, sampling was performed during peak hours (8:00–18:00) at five busiest crossroads in Zabrze. The influence of traffic-related emission contributed to the specific distribution of PAH concentrations, PAH groups, as well as diagnostic ratio values within all sampling points. The emissions from cars with diesel engine were substantial and affected the PM as well as PAH concentrations. The PAH profiles at intersections corresponded to the exhausts of cars with gasoline engines. The authors reported that at intersections, the hard coal, wood and crude oil combustion can affect the road and urban area backgrounds (traffic emission excluded), as well as the concentrations of $PM_{2.5}$ and PAHs related to PM.

Rogula-Kozłowska et al. (2014) investigated the seasonal variability of ambient mass concentrations as well as chemical composition of $PM_{2.5}$ in three locations, Katowice, Gdańsk and Diabla Góra (northern Poland, regional background site), by collecting daily the samples of $PM_{2.5}$ over a 1-year-long campaign. In comparison to the Polish permissible yearly value (25 $\mu g/m^3$), the values measured in Katowice, Gdańsk and Diabla Góra were much higher, reaching approx. 43, 50 and 75 $\mu g/m^3$, respectively. The average annual values were high as a result of the heating period and the related high monthly concentrations, especially in winter. In Katowice, Gdańsk and Diabla Góra, the yearly mass contributions to $PM_{2.5}$ amounted to 43%, 31% and 33% for carbonaceous matter, elemental carbon and organic matter, respectively. In each site, the yearly contribution of toxic metals to the mass of $PM_{2.5}$ was, on average, below 0.2%. The annual ambient concentrations of BaP in Katowice and Gdańsk were 15.4 and 3.2 ng/m^3, respectively, whereas in Diabla Góra (rural area) it nearly reached 1 ng/m^3 (the Polish annual limit of BaP). Significant variations in the component groups and concentrations of $PM_{2.5}$ occurred as a result of seasonal variations of the emissions related to the heating season. The maximum values were obtained in winter and decreased during the non-heating period.

Rozbicka and Michalak (2015) characterized the temporary as well as spatial distribution related to traffic-related air pollutants (PM_{10}, NO_2, C_6H_6 and O_3) and assessed the relation between the weather conditions and concentration of air pollution observed in three Warsaw-based monitoring stations. In all these stations, the average yearly (2011–2013) concentrations of NO_2 exceeded the limit value of 40 $\mu g/m^3$ by 114%–141%. The limit value of PM_{10} concentrations was exceeded in two stations, but the concentrations of benzene and ozone remained below the permissible value. The highest concentrations of these pollutants were observed in the station located within the city center, near communication routes. In contrast, the lowest concentrations occurred in suburbs, away from busy streets. The relation between the investigated air pollutants, ozone concentrations and weather conditions was indicated through statistical analysis.

Jędruszkiewicz et al. (2017) analyzed the variability of the PM_{10} and $PM_{2.5}$ concentrations obtained in 11 air quality monitoring stations in several urban areas in Poland (Kraków, Łódź, Poznań and Tricity). The obtained findings indicated that in Tricity, the PM_{10} and $PM_{2.5}$ threshold value was exceeded on less than 5% days. The standards in Kraków were met only in summer, whereas in winter the daily PM limit was exceeded on about 65%–90% of days. In most stations, the 10-year return level of

PM_{10} monthly maximum daily average is usually lower than 250 μg/m^3. The weather conditions in winter were not conducive to dispersion of pollutants. Czarnecka et al. (2017) performed statistical evaluation of the influence of air temperature and thermal inversion (surface and elevated) on the PM_{10}, $PM_{2.5}$ and sulfur dioxide concentrations in Warsaw, Wrocław and Gdańsk in the winter (December–February) of 2016/2017. The authors reported that the air temperature in the ground layer (up to 200 cm) significantly affected fluctuations of sulfur dioxide and both fractions of PM in each of the considered cities, especially at night (19:00–6:00). The thickness of surface inversion layers negatively affected the amount of investigated pollutants. However, substantial thickness of the daytime mixing layer, resulting from high altitude of the base of elevated inversion, exerted a highly beneficial effect on pollutant dispersion. Chlopek and Strzałkowska (2018) performed the measurements of particle mass concentration in Warsaw, within a street canyon. The 5-hour measurements preformed over a single summer day indicated relatively good correlation of the measured PM_1, $PM_{2.5}$ and PM_{10} and total particle concentrations with the results reported by a nearby air quality monitoring station. However, no significant correlations were noted with the concentrations of gaseous pollutants (carbon monoxide and nitrogen oxides) originating from vehicular traffic.

Sówka et al. (2018) reported the results from PM_{10} and $PM_{2.5}$ measurements conducted in 2010–2016 in Poznań, as well as the risk assessment analysis related to the inhalation exposure of PM_{10}-related As, Cd and Ni. The mean annual PM concentrations obtained in 2010–2016 from four measurement stations, as well as their seasonal fluctuations were investigated. The mean annual PM_{10} concentrations varied from 21 (2010) to 39 μg/m^3 (2016). The mean concentrations in the non-heating season were halved compared to the heating season. As far as heavy metals are concerned, the greatest mean seasonal concentrations equaled: 3.34 ng/m^3 for As (the heating season of 2016), 0.92 ng/m^3 for Cd (the heating season of 2012) and 10.82 ng/m^3 for Ni (the heating season of 2016). The occurrence of As in PM resulted from fossil fuel combustion in domestic fireplaces. Cd and Ni might have originated from the industry and road traffic. The greatest risk values in Poznań were obtained for As from the average concentrations during the heating seasons of 2012–2016, reaching estimated daily intake equaled 24.27×10^{-6}, 11.87×10^{-6} and 9.94×10^{-6} mg/d·kg for children, women and men, respectively.

Chambers and Podstawczyńska (2019) studied the quality of air near Łódź for the period of 2008–2011. The peak hourly concentrations of PM_{10} ($PM_{2.5}$) in the heating season were sometimes greater than 400 (300) μg/m^3, mostly due to the local domestic and traffic-related sources. The yearly average values of PM_{10} and $PM_{2.5}$ reached 33.6 and 21.1 μg/m^3, whereas the daily average values of PM_{10} exceeded the WHO limit (50 μg/m^3) 56 times per year.

Sówka et al. (2019) performed measurements and analyzed the chemical composition of PM (PM_1, $PM_{2.5}$ and PM_{10}). The authors conducted measurements in Poznań and Wrocław in 2014–2016, involving the heating and non-heating season. The greatest yearly PM concentrations were reported in 2016 reaching, on average, 41.08 (PM_1), 30.88 ($PM_{2.5}$) and 18.16 μg/m^3 (PM_{10}) in the case of Wrocław as well as 32.9 (PM_1), 30.8 ($PM_{2.5}$) and 8.5 μg/m^3 (PM_{10}) for Poznań. The performed analyses related to the chemical composition of PM indicated greater concentrations of organic and elemental carbon as well as water-soluble ions in the measurement series conducted during the heating season. The fuel combustion processes constituted the primary sources of PM emission and the impact of secondary aerosols on the quality of air was identified. In summer, migration pollutants (mineral dust) from outside of the investigated areas or from mineral and soil dust resuspension could play a significant role.

Tainio (2015) quantified the burden of disease resulting from the local transport in Warsaw. The disability-adjusted life-years (DALYs) were calculated for the air pollution related to transport, including PM, sulfur dioxide (SO_2), nitrogen oxides (NO_x), benzo(a)pyrene (BaP), nickel, lead and cadmium, as well as noise, physical activity and injuries. In urbanized environments, the health risks resulting from transport outweighed the health benefits. According to estimations, the air pollution, noise and injuries resulting from local transport in the study area were calculated to cause about 58,000 DALYs. Air pollution and noise corresponded to 44% and 46% of this burden, respectively. The physical activity related to transport was calculated as giving a health benefit amounting to 17,000 DALYs. The study reported that the transport-related health burden can be reduced by decreasing the use of motorized transport, contributing to air pollution and noise, in favor of cycling and walking.

Rogula-Kozłowska et al. (2019) studied the ambient concentrations as well as elemental composition of the particles having the aerodynamic diameter in the range of 30–108 nm (quasi-ultrafine particles, q-UFP). The considered data was acquired from 6 sites located in two large cities in Upper Silesia, i.e. Katowice and Zabrze. PM sampling was conducted at urban background site and TIs (Katowice and Zabrze), on a highway (Katowice) and an urban road (Zabrze). The ambient concentrations of q-UFP and 24 q-UFP-bound elements obtained at the aforementioned sampling sites have been discussed. The mass concentrations of q-UFP in Upper Silesia were very high, comparable to the values obtained in other areas. The percentage of total mass of the analyzed elements in the q-UFP mass indicates that the q-UFP mass in Upper Silesia comprises predominantly primary matter. For sites strongly affected by the traffic-related emissions having a low contribution of the analyzed elements to the q-UFP mass, it was suggested that probably carbonaceous matter and elemental carbon constituted the core mass of q-UFP. Most of the elements forming q-UFP were of anthropogenic origin. The impact of local PM sources on the ambient concentrations of q-UFP-bound Al, Si, S, Cl, K, Sc, Ti, V, Cd, Cr, Mn, Co and Sb was evident.

Rogula-Kozłowska et al. (2016) collected over 24-hour samples of $PM_{2.5}$ in a quasi-rural (suburban) area in Racibórz, Poland, between 1 January 2011 and 26 December 2012. The content of 28 elements was determined in the samples. During the study, the sources of $PM_{2.5}$ were indicated as well as their contributions to the total $PM_{2.5}$ concentration were determined by means of the enrichment factor (EF) analysis, principal component analysis (PCA) and multi-linear regression analysis (MLRA). In Racibórz, during the cold season of 2011–2012 (January–March and October–December), the mean ambient $PM_{2.5}$ concentration amounted to 48.7 ± 39.4 µg/m³. This value was significantly greater than at other European rural or suburban sites. Moreover, the ambient concentrations of certain toxic $PM_{2.5}$-bound elements were high as well. In the case of As, Cd and Pb, they amounted to 11.3 ± 11.5, 5.2 ± 2.5 and 34.0 ± 34.2 ng/m³, respectively. In turn, during the warm season of 2011–2012 (April–September), the concentrations of $PM_{2.5}$ and $PM_{2.5}$-bound elements were similar to those observed at other European rural (suburban) sites. The obtained findings indicate that in Racibórz, the concentrations as well elemental composition of $PM_{2.5}$ are predominantly affected by the emissions of anthropogenic origin, including biomass and coal combustion, industry and traffic.

Chambers and Podstawczyńska (2019) conducted the observations of air quality in the vicinity of Łódź in central Poland, over the period of four years. The authors reported that although the degree of industrialization has decreased since 1990, the peak hourly concentrations of PM_{10} ($PM_{2.5}$) were sometimes greater than 400 (300) $\mu g/m^3$ during the "heating season" (October – March), which was attributed to combined "imported" pollution as well as domestic and local traffic-related sources. The high emissions during the heating season resulted in the mean yearly PM_{10} and $PM_{2.5}$ values amounting to 33.6 and 21.1 $\mu g/m^3$ (exceeding the WHO guidelines by 13.6 and 11.1 $\mu g/m^3$, respectively). In turn, the mean daily PM_{10} concentration exceeded the WHO guideline of 50 $\mu g/m^3$ 56 times a year. The aforementioned authors, using the radon-based techniques for classifying the diurnal and synoptic timescale changes in atmospheric stability, indicated that the daily PM_{10} and $PM_{2.5}$ concentrations exceeded the WHO limit in two out of six mixing state categories: (1) strong persistent synoptic inversion conditions, connected with lingering anti-cyclonic systems in the non-summer months (constituting ≤15% of heating season months), and (2) fine-weather conditions conducive to the formation of strongly stable nocturnal boundary layers (15%–20% of heating season months). These mixing states were connected with low average wind speeds (1–1.5 m/s), near-surface temperature gradients of 1°C–2°C m/1, shallow nocturnal boundary layers (45–55 m), and south-easterly winds.

The WHO guideline values for daily $PM_{2.5}/PM_{10}$ were exceeded only under the conditions of strong persistent temperature inversion (PTI) (3%–15% of non-summer months) or under non-PTI stable nocturnal conditions (14%–20% of all months), co-inciding with minimum nocturnal mean wind speeds. In the non-summer months, the diurnal amplitudes of NO (CO) increased by 2–12-fold (3–7-fold) from well-mixed nocturnal conditions to PTI conditions. The highest concentrations occurred during the morning/evening commuting periods. The observations within radon-derived atmospheric mixing "class types" were analyzed to determine the relationships between weather and the air quality parameters (for instance, wind speed vs. the $PM_{2.5}$ concentration, or atmospheric mixing depth vs. the PM_{10} concentration).

4.11   METHODS AND PROSPECTS FOR REDUCING THE CONCENTRATION OF TRAFFIC-RELATED POLLUTION

Numerous actions aimed at mitigating the emission of pollutants in urban areas are mainly focused on the issues connected with urban transport. The development of transport is unavoidable from the viewpoint of economic development; however, it should be carried out in such a way that minimizes its negative impact on the natural environment and human health. Important issues related to the transport development are included in the European Transport Policy (CEC 2001) and National Transport Policy for 2006–2025 (MI 2005). The remedial actions should be simultaneously taken to: (i) prevent emission, (ii) mitigate negative effects and (iii) increase social awareness.

The issues related to education have been ignored for many years. Favorable conditions for implementation of various educational programs, popularized, by mass media, the Internet, and schools, were created only recently. Additionally, commercial organizations that monitor the air quality and make the obtained results available on-line have emerged. Special improvements in the use of public urban transport are being introduced as part of popularization of pro-ecological actions and increasing the public awareness. Moreover, actions are taken to improve the means of transport, increase the efficiency of organized urban transport, as well as improve modernization and expansion of the road infrastructure. An example of such actions includes introduction of modified fuel mixtures and fuel additives (e.g. AdBlue), improvement of combustion engines, application of efficient cleaning systems (catalytic converters, DPF, GPF), as well as popularization of hybrid and electric cars (Sówka 2017). In Lublin, as per City Hall regulation, electric and hybrid cars with CO_2 emission below 100 g/km are entitled to park for free in metered parking zones.

Development of new automotive technologies is connected with the implementation of increasingly stringent emission standards. For example, the EURO 6 standard is valid for the passenger cars manufactured after 2013 with direct-injection gasoline engines or diesel engines. This standard limits the emission of particles to the level of 4.5 mg/km and 6×10^{11} pt/km. In the case of EURO 1, which was valid from 1993 to 1996, the permissible emission of particles for diesel cars amounted to 140 mg/km.

The concentrations of traffic-related pollution can also be reduced by improving the vehicular traffic organization and expanding the restricted traffic zones. It is also important to maintain the roads in a manner limiting the secondary emission of pollutants through regular washing, repairs, and improvement of the road surfaces. The development as well as implementation of energy-saving and low-emission solutions in public transport should be a priority.

Beltways that take over the transit traffic, appropriate selection of intercity roads, and implementation of grade-separated intersections facilitating traffic and reducing the emission of pollutants are essential elements of the road infrastructure development. Apart from that, bus lanes are being introduced in numerous cities, which are restricted to buses and taxis. In the case of Lublin, the first, northeastern part of the beltway was built in 2011–2014, whereas the second, western part was built in 2014–2016. The cycling lanes built in the city also encourage using bicycles instead of cars. In 2016, there was about 130 km of bicycle routes in Lublin. Additionally, a well-developed bicycle rental infrastructure was created in the city.

4.12 CONCLUSIONS

The presented results of traffic-related pollution measurements in Lublin confirm their significant impact on the quality of air in Lublin and are in agreement with the values found in literature and are subject to similar dependences.

The results obtained in the study confirm that the variations in concentrations of particle number and mass of aerosols as well as particle size distributions along the investigated road and its proximity resulting from the traffic conditions can largely contribute to the exposure as well as negative health effects for pedestrians and commuters.

The findings of the research indicate that the particle mass and number concentrations found on the monitored route are dependent upon the traffic intensity. In turn, the resultant exposure of commuters and pedestrians is related to the time of the day and route section.

Further research should involve determining the relation between particle concentrations as well as the exposure under different meteorological conditions in different seasons, enabling to develop effective methods for health risks elimination.

The obtained results proved that the traffic conditions, which differ throughout the day and in particular seasons, directly affect the traffic-related particle number and mass concentrations obtained on the sidewalk along the investigated route. The traffic conditions significantly influence the exposure of pedestrians and can be important when considering the health benefits stemming from active travel as well as the detrimental effect from increased air pollution intake. In order to remedy the situation, it is necessary to increase the public awareness on the health hazards resulting from traffic-related pollution. The obtained findings show that the lowest exposure in vicinity of the monitored road occurs in the summertime evenings. The results reported in the study may be of use during urban development and while re-designing or building new sidewalks away from busy roads.

Further research is needed in order to establish the relations between the daily and seasonal changes in traffic intensity, in addition to the characteristics of roads and vehicles in various parts of Lublin, as well as the particle concentration levels along the road and sidewalk, with the potential impact on the health of pedestrians and commuters, resulting from the exposure to particles. The study should also account for the actions taken in Lublin to reduce the particle exposure.

Effective methods of mitigating the health risks could be developed by conducting simultaneous, multi-point and extended research to determine the relation between particle concentrations and exposure of pedestrians under different meteorological conditions in particular seasons.

Chapter 5

Health effects

Increased exposure to outdoor air pollutants can have very serious health effects (Valavanidis et al. 2008). According to the data contained in the "Air quality in Europe – 2017 report", aerosol particles and NO_2 are considered the most serious pollutants having a negative impact on human health. It is estimated that in 2014 throughout Europe, long-term exposure to $PM_{2.5}$ particles was responsible for approx. 428,000 premature deaths, of which in Poland for over 46,000. The indicator of the total number of life years lost due to exposure to $PM_{2.5}$ particles in our country reached the level of 1,455/100,000 people. In the case of NO_2, this indicator amounted to 54/100,000 people, and the estimated number of premature deaths caused by exposure to NO_2 reached the value of 1700 (EEA 2017). Both aerosol particles, nitrogen oxides, carbon monoxide and volatile organic compounds, which are the main pollutants emitted from transport, significantly increase the risk of serious respiratory diseases, including cancers of the lung, pharynx and larynx; chronic obstructive pulmonary disease (COPD); asthma; and allergies (Gładka and Zatoński 2016). Studies conducted in Warsaw and Silesia in order to estimate the risk of respiratory problems among people exposed to increased exposure to traffic pollution showed a significant increase in airflow disorders through the bronchi (Badyda 2010). For residents of areas located on busy streets (compared to non-urbanized areas), this increase was almost threefold, and among non-smokers, it was more than fourfold. Rogula-Kozłowska et al. (2008), on the basis of research conducted in Zabrze, estimated that increased concentrations of PM_{10} and $PM_{2.5}$ particles cause a 4%–10% increase in upper respiratory tract diseases among people living at street crossings. On the other hand, according to studies conducted in Gliwice, increased concentrations of PM_{10} particles in street canyons may cause a 10% increase in morbidity related to the functioning of the respiratory system (Grynkiewicz Bylina et al. 2005). In Krakow, which is considered the most polluted large city in the EU, the number of deaths attributed to poor air quality caused, inter alia, by transport pollution, is estimated at several hundred people per year (Badyda et al. 2016).

Cardiovascular diseases – heart attack, arterial hypertension, atherosclerosis and heart failure (Lee et al. 2014, Du et al. 2016) – are also frequently observed effects of exposure to traffic pollution. Numerous scientific reports also indicate a relationship between increased exposure to traffic pollution and a multiplied risk of stroke and nervous system ailments, including problems with memory and concentration, depression and a higher risk of developing Alzheimer's disease (Moulton and Yang 2012, Walton 2018). Research conducted in London on the exposure of users of various means of transport showed that people traveling by car or bus are 1.5 times more

DOI: 10.1201/9781003206149-5

exposed to aerosol particles than people walking on foot (Kaur et al. 2005). From the research of Fruin et al. (2008) carried out in the United States, it was found that approx. 6% of the time spent in the car translates into approximately 36% of the daily exposure to aerosol particles. In turn, Dons et al. (2012) showed that in Belgium about 4% of the time spent in the car is responsible for almost 15% of the daily exposure. Similar results are reported for other European Union countries (Knibbs et al. 2011).

Short-term exposure to relatively high concentrations of aerosol pollutants in the air may cause numerous disease symptoms, especially among the elderly and the sick, children and pregnant women (GIOS 2017). Conversely, long-term exposure to elevated concentrations of aerosol pollutants increases mortality, which translates into shorter life expectancy. Pope et al. (2009) showed that reducing long-term exposure to $PM_{2.5}$ by 10 µg/m^3 increases life expectancy by over half a year (0.61 ± 0.20 years). In the case of places with increased concentrations of these pollutants (and in such a situation is Lublin), they can shorten the lives of inhabitants by over a year (Ballester et al. 2008).

Actions to reduce air pollution concentrations of traffic pollution should be one of the most important and urgent challenges in the field of public health in Lublin and the surrounding region. Increased morbidity and incidence of major diseases, including cancer, may be related to exposure to traffic pollution. According to the Statistical Guide for Healthcare of the Lubelskie Voivodeship for 2013, 30,667 people suffered from cancer in the group of adults, and the number of new cases was 6,097. The prevalence and incidence of cancer were, respectively, 181.4/10,000 population and 36.1/10,000 population. The most common causes of cancer deaths, regardless of gender, were bronchial and lung cancers. Depending on gender, the incidence rate of malignant neoplasms of the bronchi and lungs was 85.8/100,000 men and 27.4/100,000 women. In Lublin, in 2013, among people over 18 years of age, the number of cancer patients was 6,778 people, of which 1,081 were new cases. In terms of the number of inhabitants, the prevalence and incidence rates of cancer were 213.6/10,000 population and 34.1/10,000 population.

Transport pollution has negative consequences not only for people but also for other elements of the ecosystem. They are precursors of acidifying compounds in the environment, and their presence may have a negative effect on vegetation and animals but also on inanimate matter, e.g. building structures. Moreover, the content of greenhouse gases such as carbon dioxide, methane, nitrous oxide, ozone, carbon monoxide, as well as nitrogen oxides and volatile organic compounds may also significantly intensify the greenhouse effect (Badyda 2010).

Nitric oxide comes mainly from road traffic because it is emitted by car engines (Hesterberg et al. 2009, Richmont-Bryant et al. 2017). It is primarily irritating to the respiratory tract. It penetrates the lungs causing respiratory diseases, shortness of breath, coughing, wheezing, bronchospasm and, at high concentrations, even pulmonary edema. Concentrations above 0.2 ppm are believed to cause adverse health effects in humans; higher than 2.0 ppm affects lymphocytes and triggers the immune response. High nitrogen dioxide levels can contribute to chronic lung disease. Long-term exposure to NO_2 weakens the sense of smell. Symptoms not related to respiratory symptoms such as irritation of the eyes, throat and nose have also been reported (Chen et al. 2007).

High levels of nitrogen dioxide are not only harmful to humans but also negatively affect vegetation. It has been observed that they reduce yields and plant growth efficiency. In addition, NO_2 can discolor fabrics and reduce visibility.

Sulfur dioxide is emitted mainly from the combustion of fossil fuels or from industrial activities. The annual SO_2 standard is 0.03 ppm (US EPA 2019). Too high concentration of sulfur dioxide in the air adversely affects the life and health of humans, animals and plants. The most vulnerable people are those with low immunity, those with lung diseases, the elderly and children. Health problems related to sulfur dioxide emissions are recorded mainly in industrialized areas. The main complaints are irritation of the respiratory tract, bronchitis and bronchospasm. They result from the properties of sulfur dioxide as a substance irritating the sensory organs and penetrating deep into the lungs, where it is converted into bisulfite, it interacts with the sensory receptors, causing the bronchospasm mentioned above. In addition, there is reddening of the skin, damage to the eyes (tearing and clouding of the cornea) and mucous membranes, and deterioration of health in people with cardiovascular diseases (Chen et al. 2007). In addition, sulfur dioxide emissions are believed to have a negative impact on the environment and cause soil acidification and acid rain (WHO 2000).

Another air pollutant is lead – a heavy metal produced by some industrial plants and emitted from certain gasoline engines (land and air transport), batteries, coolers, waste and sewage incineration plants (Pruss-Ustun et al.).

Lead poisoning is currently a major threat to public health, especially in developing countries, due to its harmful effects on living organisms and the environment.

Lead can enter the body by inhalation, ingestion, or skin absorption. Lead transport through the placenta has also been reported – the younger the fetus, the more severe the toxic effects. It affects the nervous system of the fetus. Among other things, brain edema has been observed. When lead is deposited by inhalation, it accumulates in the blood, soft tissues, lungs, liver, bones and nervous, circulatory and reproductive systems. In adults, loss of concentration, memory, and muscle and joint pain has been reported (Goyer 1990, NIH 2013).

Children and newborns are extremely susceptible. Minimal doses of lead, which is considered to be neurotoxic, can cause learning disabilities, memory impairment, hyperactivity and mental retardation (Farhat et al. 2013).

Increased amounts of lead in the environment are harmful not only to organisms but also to plants, including crops. High lead levels are associated with neurological effects in vertebrates and animals (Assi et al. 2016).

The presence of PAHs in the environment is common because the atmosphere is where they spread. They occur in coal and tar sediments and are formed in the processes of incomplete combustion of organic matter, during forest fires, or as a result of road traffic. PAH compounds (benzopyrene, acenaphthylene, anthracene and fluoranthene) are considered to be toxic and mutagenic and carcinogenic substances. They are considered to be important factors in the formation of lung cancer (Abdel-Shafy and Mansour 2016).

It has also been found that volatile organic compounds (VOCs) – toluene, benzene, ethylbenzene and xylene – are the cause of cancer in humans. The introduction of new products and materials into rooms increases the concentration of VOCs, which results in short- and long-term adverse health effects. VOCs are responsible for odors in indoor air. Short-term exposure causes irritation of the eyes, nose, throat and mucous membranes, while long-term exposure involves toxic reactions. The assessment of the toxic effects of complex VOC mixtures is difficult to quantify. These pollutants may be synergistic, antagonistic, or neutral (Molhave et al. 2004, Ebersviller et al. 2012).

Dioxins are produced in industrial and natural processes, including forest fires and volcanic eruptions. They accumulate mainly in food – meat (especially adipose tissue of animals), dairy products, fish and crustaceans. Short-term exposure to high concentrations of dioxins can cause skin lesions. Long-term exposure may lead to developmental problems; impaired immune, endocrine and nervous systems; reproductive sterility; and even cancer (WHO 2010).

Undoubtedly, the combustion of fossil fuels is responsible for a significant part of air pollution. This pollution can be anthropogenic (agriculture, industry, or transport), but it can also come from natural sources. Interestingly, the air quality standards set out in the European Air Quality Directive are less stringent than the WHO guidelines.

Despite the significant amount of scientific research to identify the sources of emissions and their health effects, there is currently insufficient evidence to conclusively attribute specific health effects to specific sources. This is due to the lack of markers characteristic of the source and the contamination itself. Identification is also hindered by the fact that in the case of PM, the toxic composition and its size are also important (WHO 2007, Stanek et al. 2011). Therefore, the impact of PM on health is the subject of intense research by scientists around the world. Small particles penetrate the deeper parts of the lungs and settle there. Figure 5.1 shows the deposition of particles in the respiratory system depending on their size corresponding to Andersen impactor stages.

It is already known that air pollution is a major cause of human morbidity and mortality and increases the risk of diseases, including respiratory diseases (Huang et al. 2016, Feng et al. 2016, Li et al. 2015, Gao et al. 2018, Polichetti et al. 2009, Janssen et al. 2013,

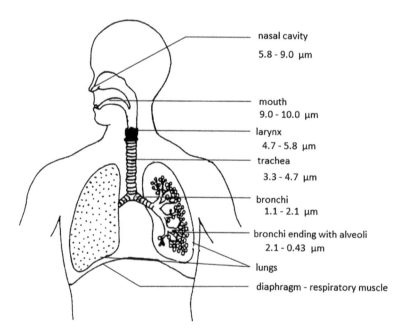

nasal cavity
5.8 - 9.0 μm

mouth
9.0 - 10.0 μm

larynx
4.7 - 5.8 μm

trachea
3.3 - 4.7 μm

bronchi
1.1 - 2.1 μm

bronchi ending with alveoli
2.1 - 0.43 μm

lungs

diaphragm - respiratory muscle

Figure 5.1 Particulate matter size and penetration in respiratory tract. (Based on Nave & Mueller 2013).

Brunekreef and Holgate 2002, Polichetti et al. 2013, Nhung et al. 2018). Exposure to air pollution affects not only the respiratory system (including asthma, lung cancer) but also the cardiovascular and nervous systems.

The consequences affect only some of the people living in a given area and vary in intensity, usually caused by the state of health and individual sensitivity. It is also associated with genetics, but many data confirm that organisms in development and the elderly are particularly sensitive to pollution from exhaust gases. The size of the risk in children depends on the anatomical and physiological specificity:

- shorter and narrower airways,
- frequent breathing through the mouth,
- ineffective filtration of particles in the nasal passages,
- greater minute ventilation per unit of body weight,
- immature detoxification systems,
- immaturity of the immune system,
- the respiratory system is still under development (up to 3–8 years of age).

$PM_{2.5}$ particulate matter is estimated to be responsible for 0.5 million premature deaths in Europe (less than 400,000 in the 28 EU countries), including almost 80% of deaths from respiratory disease and lung cancer (EEA 2015). These particles are so small that they can penetrate deep into the lungs and into the bloodstream. It has been proven that exposure to $PM_{2.5}$ over many years can have adverse health effects. $PM_{2.5}$ fine particles pose a greater risk to health due to the place of deposition in the respiratory system (Kelishadi and Poursafa 2010, Zhang et al. 2019) (Figure 5.1).

Correlation has also been demonstrated between short- and long-term $PM_{2.5}$ exposure and acute nasopharyngitis (Zhang et al. 2019) and between long-term PM exposure and cardiovascular disease and infant mortality. The occurrence of PM_{10} and $PM_{2.5}$ particles in the atmosphere is prolonged due to their small size, which allows them to be suspended for a long time in the atmosphere and transported and spread to distant places (Wilson and Suh 1997). Studies have shown an association between increased hospital admissions and deaths from heart or lung disease and exposure to particles. Despite numerous studies, there is currently no threshold below which exposure to PM would not cause any health effects. Health effects may occur after short- and long-term exposures. Short-term exposure is believed to exacerbate existing disease, while long-term exposure likely causes disease and increases the rate of progression. It is estimated that exposure to $PM_{2.5}$ shortens life expectancy by an average of about 8.6 months. According to the morbidity data, scientists call $PM_{2.5}$ one of the six greatest risk factors for death in the world, causing more than 4 million deaths each year. With the addition of other pollutants from exhaust fumes, that number approaches 7 million deaths each year.

Studies have shown that $PM_{2.5}$ components can induce oxidation of lung cells, which may be the main cause of cell damage (Donaldson and Beswick 1996, Greenwell et al. 2002, Rahman and Macnee 1996, Kelly 2003). In 1996, Donaldson and Beswick reported that the surface of the particles alone can generate free radicals. On the surface of $PM_{2.5}$, iron, copper, zinc, manganese and other elements were detected as well as polycyclic aromatic hydrocarbons. These ingredients can increase free radical production, consume antioxidant ingredients and cause oxidative stress (Donaldson and Beswick 1996).

Many studies (Kim et al. 1997, Valavanidis et al. 2005, Xing et al. 2011) have confirmed that reactive oxygen species generated by molecules, especially those soluble in water, produce hydroxyl radicals, which are the main factor in oxidative damage to DNA. Damaged DNA, not repaired on time, can cause irreparable damage. Mehta et al. (2008) discovered that the molecules could not only damage DNA and inhibit DNA repair but could also promote the replication of damaged DNA fragments. It is believed that each 10 $\mu g/m^3$ increase in PM concentration increases the respiratory system morbidity by 1%. Esposito et al. (2014) observed that an increase in PM_{10} levels increases the risk of pneumonia by about 8% in children with a history of wheezing or asthma.

Most of the literature on air pollution-related mortality (due to $PM_{2.5}$) has emerged from studies in HICs (high-income countries) like the USA, France, Canada and Scandinavian countries (Table 5.1).

Table 5.1 Mortality due to $PM_{2.5}$ in different countries (confidence interval)

Country	Mortality due to $PM_{2.5}$	References
USA	The mortality hazard ratios per 10 $\mu g/m^3$ long-term exposure to $PM_{2.5}$ for the time period of 1999–2015, were 1.12 for all-cause mortality, 1.23 for cardiopulmonary mortality, and 1.12 for lung cancer (LC) mortality.	Pope et al. (2019)
	Death due to cardiovascular disease (CVD) (56,070.1 deaths), cerebrovascular disease (40,466.1 deaths), chronic kidney disease (7175.2 deaths), chronic obstructive pulmonary disease (COPD) (645.7 deaths), dementia (19,851.5 deaths), type 2 diabetes mellitus (T2DM) (501.3 deaths), hypertension (30,696.9 deaths), LC (17,545.3 deaths) and pneumonia (8,854.9 deaths) for the year 2006.	Bowe et al. (2019)
	(2000–2009) $PM_{2.5}$ exposure was significantly associated with total mortality (hazard ratio HR = 1.03) and CVD mortality (HR = 1.10), association with respiratory mortality was not statistically significant (HR = 1.05). The sources of long-term $PM_{2.5}$ exposures, fossil fuel combustion, especially coal burning (HR = 1.05) and diesel traffic (HR = 1.03), was associated with the increases in ischemic heart disease (IHD) mortality in nationwide population spread across 100 US metropolitan areas (2000–2005).	Thurston et al. (2016a)
Canada	Circulatory disease (HR = 1.19), respiratory disease mortality (HR = 1.52), CVD (HR = 1.25), cardiometabolic disease (HR = 1.27), IHD (HR = 1.36), COPD mortality (HR = 1.24) 1998–2011.	Pinault et al. (2016, 2017)
Europe	Long-term exposures $PM_{2.5}$ – an increase in mortality (HR = 1.09; France 1989 - 2008). Denmark 49,564 death (CVD, 1993–2015) (HR = 1.29). 2009–2014, about 299 deaths per year (3 are infants) were estimated to be caused by $PM_{2.5}$ in Verona, Italy.	Bentayeb et al. (2015), Hvidtfeldt et al. (2019), Burnett et al. (2014), Pozzer et al. (2019), Lanzinger et al. (2016)

(Continued)

Table 5.1 (Continued) Mortality due to PM$_{2.5}$ in different countries (confidence interval)

Country	Mortality due to PM$_{2.5}$	References
Korea	For an average PM$_{2.5}$ concentration of 30.2 µg/m^3 (1990–2013), the estimated number of premature deaths was 17,203, ischemic stroke (5,382), cancer of trachea, bronchus and lung (4958), hemorrhagic stroke (3,452) and IHD (3,432), which accounted for 6.4% of all deaths in Korea.	Kim et al. (2018)
China	In Beijing, for the year 2014, 3117 annual deaths were due to lung cancer, 6318 annual deaths were due to cardiovascular, and 1697 annual deaths were due to respiratory diseases. Every 10 µg/m^3 increase in daily average PM$_{2.5}$ concentration corresponded to 0.26%–0.35% increase in daily mortality in Shanghai (2012–2014). The results from a study showed that 2.19 million (2013), 1.94 million (2014) and 1.65 million (2015) premature deaths were attributed to PM$_{2.5}$ long-term exposure in China.	Li et al. (2017), Fang et al. (2017), Chen et al. (2017a), He et al. (2016), Li et al. (2017, 2018), Lin et al. (2016)
Brazil	About 89% of the annual PM$_{2.5}$ concentration analyzed were higher than WHO guideline (10 µg/m^3) (2000–2017 Sao Paulo).	Andreão et al. (2018)
Thailand	PM$_{2.5}$ was found to have an association with COPD, coronary artery disease (CAD) death, and sepsis in Chiang Mai, Thailand.	Pothirat et al. (2019)
Iran	PM$_{2.5}$ had the strongest relationship with respiratory deaths HR = 1.06 in Tehran, Iran, during the time period of 2005–2014.	Dehghan et al. (2018)
Egypt	Approximately 11% of the mortality in the population older than 30 years in Cairo, Egypt, can still be attributed to PM$_{2.5}$, which corresponds to 12,520 yearly premature deaths.	Burnett et al. (2014), Wheida et al. (2018)
India	2016 – the attributable burden of disease due to the PM$_{2.5}$ exposure for India was 590.5, which was less than China but more than the USA. Premature mortality in India in 2015 due to the PM$_{2.5}$ exposure was estimated to be the highest due to cerebrovascular disease (CEVD) (0.44 million), followed by IHD (0.40 million), chronic obstructive pulmonary disease (COPD) (0.18 million) and lung cancer (LC) (0.01 million), with a total of 1.04 million deaths.	Bowe et al. (2018), Guo et al. (2018)

Increased levels of PM$_{2.5}$ are associated with high rates of respiratory morbidity (Table 5.2).

The association between PM$_{2.5}$ and cardiovascular morbidity concerns diseases like cardiovascular hospitalizations, myocardial infarction, CAD/IHD, hypertension, arrhythmia and heart failure is presented in Table 5.3 and Table 5.4.

Diabetes incidence was found to increase along with the PM$_{2.5}$ concentrations in the countries of different income groups. India specifically has significant association between diabetes and the PM$_{2.5}$ concentrations (Table 5.5).

Table 5.2 Respiratory morbidity due to PM$_{2.5}$ in different countries

Country	Respiratory morbidity due to PM$_{2.5}$	References
USA	A 4% increase in the number of asthma cases between 2007 and 2012, a 7.2% increase in the daily admissions for the ER for asthma patients, exposure to levels greater than 0 µg/m³ resulted in an increase risk indicators for all causes and respiratory exposure by 15% and 21%, respectively (2002–2010).	Castner et al. (2018), Chang et al. (2019), Makar et al. (2017)
Canada	The relative risk of lung cancer was 1.09 for 10-µg/m³, accounting for 1,739 lung cancer cases in 2015.	Gogna et al. (2019)
Europe	The analysis of data from Warsaw, Białystok, Bielsko-Biała, Krakow and Gdańsk (Poland) in the years 2014–2017 showed an increase by 0.9%–4.5% in hospitalizations due to respiratory system diseases. In Italy (2001–2008) for lung cancer (HR = 1.18) and lower respiratory tract infections (HR = 1.07). In Spain, for increased risk of respiratory diseases by 5.1% and increase in admissions by 7.7% in emergency cases. In Las Palmas de Gran Canaria, there is a positive association between PM$_{2.5}$ and a higher asthma hospitalization unit (21.9%).	Warkentin et al. (2019), Slama et al. (2019), Gandini et al. (2018), López-Villarrubia et al. (2016)
China	The average concentration of PM$_{2.5}$ in China (Shenzhen) in 2014–2018 was 44 µg/m³. During the same period, the risk of hospital visits for respiratory diseases was determined to be 5.40, indicating a link between daily exposure to particulate matter and hospital visits for respiratory diseases. The number of emergency room visits increased by 0.72% due to the 10 µg/m³ increase in PM$_{2.5}$ between March 1, 2015 and February 28, 2018. 0.168% increase in daily outpatient visits for rare diseases in Jinan, China, between January 2012 and December 2016. Research results in Shijiazhuang, China, showed 13% hospitalizations (January 1, 2013 to December 31, 2016) related to with exposure to PM$_{2.5}$.	Qu et al. (2019), Zhang et al. (2019b), Wang et al. (2018), Chen et al. (2019)
Brazil	In Volta Redonda (Paraiba Valley, Brazil), the 2012 mean concentration of PM$_{2.5}$ was 17.2 µg/m³, and the effects of exposure to PM$_{2.5}$ were significant (1.017): acute bronchitis and bronchiolitis (1.022) and asthma (1.020).	Nascimento et al. (2016)
Iran	Increasing the concentration of PM$_{2.5}$ by 10 µg/m³ corresponded (1.03) to the increase in the number of hospitalizations due to respiratory diseases with average daily pollutant concentrations of 24.30 µg/m³ for PM$_{2.5}$ in 2010–2015 in Arak, Iran.	Vahedian et al. (2017)
India	Approximately 5.12% of asthma cases were attributed to long-term PM$_{2.5}$ exposure. Relative risk studies (2015–2016) available from various studies indicate a higher risk of COPD and lung cancer (LC) in northern (COPD = 1.35, LC = 1.50) and eastern cities (COPD = 1.27, LC = 1.38) which have higher PM$_{2.5}$ levels compared to southern or western cities.	Ai et al. (2019), Burnett et al. (2014), Sahu et al. (2019)

Table 5.3 Cardiovascular morbidity due to $PM_{2.5}$ in different countries

Country	Cardiovascular morbidity due to $PM_{2.5}$	References
USA	Increasing exposure from levels lower than 8 µg/m^3 to levels higher than 8 µg/m^3 increased the risk of hospitalization due to cardiovascular complaints by 18% (2002–2010). Based on a study in Atlanta (Georgia), it has been estimated that $PM_{2.5}$ components, especially water-soluble iron, have the strongest effect on cardiovascular problems (1.012). In the years 2002–2009, an increase in the mean annual concentration of $PM_{2.5}$ by 1 µg/m was associated with an 11.1% relative increase in the probability of developing CAD and a 14.2% increase in the probability of a heart attack – North Carolina, USA. The average concentration of $PM_{2.5}$ during this period was 12.4 µg/m^3.	Makar et al. (2017), Ye et al. (2018), McGuinn et al. (2016)
Europe	A 5 µg/m^3 increase in $PM_{2.5}$ is related to a risk factor of 1.13 for coronary heart disease in Europe (Finland, Sweden, Denmark, Germany and Italy in 2008 and 2011). The studies concerned cardiovascular diseases (1.05) and heart attack (1.15) in Italy in 2001–2008. An increase in $PM_{2.5}$ concentration by 10 µg/m^3 was associated with a higher risk of admission to hospital due to acute myocardial infarction (1.32) among the inhabitants of Krakow in 2012–2015.	Wolf et al. (2015), Gandini et al. (2018), Konduracka et al. (2019)
China	An increase in $PM_{2.5}$ concentration by 10 µg/m^3 could increase the risk of hypertension by 11% (1.11) in the years 2004–2015. In Shanghai, China (2013–2014), an increase in the number of hospitalizations due to IHD (ischemic heart disease) was estimated by 0.25% with an increase in $PM_{2.5}$ concentration by 10 µg/m^3. $PM_{2.5}$ was associated with IHD admissions in 2014–2015 and resulted in a 1.7% increase in hospital admissions in 26 major Chinese cities. From 2014 to 2015, China saw an increase of 2.09% in hospital admissions for cardiac arrhythmias, and the mean concentration of $PM_{2.5}$ during this period was 63.5 µg/m^3. In Nanjing, China, a 10 µg/m^3 increase in $PM_{2.5}$ was associated with a 0.42% increase in the incidence of cardiovascular disease (CCD). The mean daily $PM_{2.5}$ concentration in Shanghai (55 µg/m^3) was associated with a 1.7% increase in deaths from cerebrovascular disease.	Xu et al. (2017), Huang et al. (2019), Dai et al. (2018), Wang et al. (2019), Leepe et al. (2019), Huang et al. (2018)
B4razil	The exposure to $PM_{2.5}$ (ranging from 8.5 to 89.7 µg/m^3) was significantly associated with an increase in average blood pressure in Sao Paulo, Brazil, for the time period of 2010–2012.	Santos et al. (2019)
India	Studies conducted in eight cities in India in 2015–2016 indicated a higher risk of developing IHD (ischemic heart disease) in northern (1.39) and eastern (1.35) cities compared to southern or western cities.	Burnett et al. (2014), Sahu et al. (2019)

Table 5.4 Stroke/cerebrovascular morbidity due to $PM_{2.5}$ in different countries

Country	Stroke/Cerebrovascular Morbidity to $PM_{2.5}$	References
Europe	An increase in $PM_{2.5}$ concentration by 10 µg/L increased the incidence of intracerebral hemorrhage by an average of 5.7% (2010–2015 Algarve, Portugal). Similarly, for stroke for almost 5,000 people in the cities of Bochum, Essen and Mulheim / Ruhr in Germany, the HR was 3.20. A Swedish study using data from Gothenburg, Stockholm and Umea between 1990 and 2011 showed that $PM_{2.5}$ (range 2.9–22 µg/m³) was not associated with stroke.	Hoffmann et al. (2015), Nzwalo et al. (2019), Ljungman et al. (2019)
Korea	Amyotrophic lateral sclerosis (ALS) was significantly associated with an increase in $PM_{2.5}$ (1.21) among 617 patients who visited the HED in Seoul with ALS in 2008–2014.	Myung et al. (2019)
China	There was an increase of 0.31% in daily admissions for ischemic stroke for each 10 µg/m³ increase in $PM_{2.5}$ in Beijing, China, over the period (2010–2012). In 2014–2016, an increase in $PM_{10-2.5}$ concentration by 10 µg/m³ was associated with an increase of 0.91% in hospital admissions for ischemic stroke in 172 Chinese cities.	Tian et al. (2019), Li et al. (2019)
India	Between 2015 and 2016, RR estimates available from various studies showed a greater risk of stroke in northern (RR = 2.06) and eastern (RR = 1.93) cities with higher $PM_{2.5}$ levels than in southern cities (RR = 1.54) or Western countries (RR = 1.59). It has been observed that the risk of stroke is higher with cardiometabolic and respiratory diseases in cities of northern India 1.37–1.52, southern India 1.20–1.31, eastern India 1.40–1.52 and western India 1.24–1.35.	Burnet et al. (2014), Sahu et al. (2019)

Table 5.5 Diabetes morbidity due to $PM_{2.5}$ in different countries

Country	Diabetes morbidity due to $PM_{2.5}$	References
Canada	In 2007–2014, an increase in $PM_{2.5}$ by 10 µg/m³ was associated with an increased risk of diabetes by 5.34%.	Requia et al. (2017)
Europe	An average $PM_{2.5}$ – 19.6 µg/m³ (2008–2013). For every 5 µg/m³ increase in the $PM_{2.5}$ concentration, the hazard ratio was 1.002.	Renzi et al. (2018)
Australia	Daily $PM_{2.5}$ concentrations (2009–2014) revealed that elevated $PM_{2.5}$ was associated with hypoglycemia (1.07) and fainting (1.09) per 10-µg/m³ increase in daily $PM_{2.5}$.	Johnston et al. (2019)
China	With a 10 µg/m³ increase in $PM_{2.5}$ concentration with long-term exposure, there was an increase in the incidence of diabetes by 15.66% between 2004 and 2015 for 27,840 Chinese adults from 15 provinces in China. A 0.53% increase in hospitalizations due to a 10 µg/m³ increase in $PM_{2.5}$ concentration was estimated in 2014–2016 in Shijiazhuang, China, where the mean $PM_{2.5}$ concentration in the study period was 105 µg/m³.	Liang et al. (2019), Song et al. (2018)
India	The disease burden related to exposure to $PM_{2.5}$ in India is about 45%.	Bowe et al. (2019)

The principal findings are as follows:

- High risk of mortality and morbidity due to respiratory, cardiometabolic and cerebrovascular disorders occur even at low $PM_{2.5}$ concentrations.
- Increase in concentrations in the countries where the $PM_{2.5}$ concentrations are low will have greater health effects than increases in the countries with high $PM_{2.5}$ concentrations.

References

Abdel-Shafy, H.I. & Mansour, M.S.M. 2016. A review on polycyclic aromatic hydrocarbons: source, environmental impact, effect on human health and remediation. *Egypt J Pet* 25: 107–123.

Adamiec, E., Jarosz-Krzemińska, E. & Wieszała, R. 2016. Heavy metals from non-exhaust vehicle emissions in urban and motorway road dusts. *Environ Monit Assess* 188: 369.

Adeniran, J.A., Yusuf, R.O. & Olajire, A.A. 2017. Exposure to coarse and fine particulate matter at and around major intra-urban traffic intersections of Ilorin metropolis, Nigeria. *Atmos Environ* 166: 383–392.

Ai, S., Wang, C., Qian, Z. et al. 2019. Hourly associations between ambient air pollution and emergency ambulance calls in one central Chinese city: Implications for hourly air quality standards. *Sci Total Environ* 696: 133956.

Alharbi, B., Shareef, M.M. & Husain, T. 2015. Study of chemical characteristics of particulate matter concentrations in Riyadh, Saudi Arabia. *Atmos Pollut Res* 6 (1): 88–98.

Amato, F., Pandolfi, M., Alastuey, A., Lozano, A., Contreras González, J. & Querol, X. 2013. Impact of traffic intensity and pavement aggregate size on road dust particles loading. *Atmos Environ* 77: 711–717.

Andreão, W.L., Albuquerque, T.T.A. & Kumar, P. 2018. Excess deaths associated with fine particulate matter in Brazilian cities. *Atmos Environ* 194: 71–81.

Arkouli, M., Ulke, A.B., Endlicher, W., Baumbach, G., Schultz, E., Vogt, U., Müller, M., Dawidowski, L., Faggi, A., Wolf-Benning, U. & Scheffknecht, G. 2010. Distribution and temporal behavior of particulate matter over the urban area of Buenos Aires. *Atmos Pollut Res.* 1: 1–8.

Arya, S.P. 1999. *Air Pollution Meteorology and Dispersion.* New York: Oxford University Press.

Assi, M.A., Hezmee, M.N.M, Haron, A.W. et al. 2016. The detrimental effects of lead on human and animal health. *Vet World* 9: 660–671.

Ault, A.P., Moore, M.J., Furutani, H. & Prather, K.A. 2009. Impact of emissions from the Los Angeles port region on San Diego air quality during regional transport events. *Environ Sci Technol* 43: 3500–3506.

Badyda, A.J. 2010. Environmental threats from transport (in Polish). Nauka 4: 115–125.

Ballester, F., Medina, S., Boldo, E., Goodman, P., Neuberger, M., Iniguez, C. & Künzli, N. 2008. Reducing ambient levels of fine particulates could substantially improve health: a mortality impact assessment for 26 European cities. J Epidemiol Commun H 62: 98–105.

Banks, R.F., Tiana-Alsina, J., Rocadenbosch, F. & Baldasano, J.M. 2015. Performance evaluation of the boundary-layer height from lidar and the weather research and forecasting model at an urban coastal site in the north-east Iberian Peninsula. *Bound-Layer Meteorol* 157: 265–292.

Bauwens, M., Compernolle, S., Stavrakou, T., Müller, J.F., van Gent, J., Eskes, H., Levelt, P.F., van der A, R., Veefkind, J.P., Vlietinck, J., Yu, H. & Zehner, C. 2020. Impact of coronavirus outbreak on NO$_2$ pollution assessed using TROPOMI and OMI observations. *Geophys Res Lett* 47: e2020GL087978.

Bekbulat, B., Apte, J.S., Millet, D.B., Robinson, A., Wells, K.C. & Marshall, J.D. 2020. $PM_{2.5}$ and ozone air pollution levels have not dropped consistently across the US following societal COVID response. *ChemRxiv* (preprint). Doi: 10.26434/chemrxiv.12275603.v7.

Bentayeb, M., Wagner, V., Stempfelet, M. et al. 2015. Association between long-term exposure to air pollution and mortality in France: A 25-year follow-up study. *Environ Int* 85: 5–14.

BIP (Bulletin of Public Information of the Lublin City Council) 2017. Number of vehicles registered in Lublin in years 2000–2016 (in Polish), https://www.bip.gov.pl, (Accessed in December 2017).

Dirmill, W., Tomsche, L., Sonntag, A., Opelt, C., Weinhold, K., Nordmann, S. & Schmidt, W. 2013. Variability of aerosol particles in the urban atmosphere of Dresden (Germany): effects of spatial scale and particle size. Meteorol Z 22: 195–211.

Bleecker, M.L. 2015. Chapter 12- Carbon monoxide intoxication, Editor(s): Marcello Lotti, Margit L. Bleecker, *Handbook of Clinical Neurology*, Elsevier, Vol. 131, pp. 191–203, ISSN 0072-9752, ISBN 9780444626271.

Bliss, B., Tran, K.I., Sioutas, C. & Campbell, A. 2018. Ambient ultrafine particles activate human monocytes: Effect of dose, differentiation state and age of donors. *Environ Res* 161: 314–320.

Blumenthal I. 2001. Carbon monoxide poisoning. *J Royal Soc Med* 94(6): 270–272.

Bowe, B., Xie, Y., Li, T., Yan, Y., Xian, H. & Al-Aly, Z. 2018. The 2016 global and national burden of diabetes mellitus attributable to $PM_{2.5}$ air pollution. *Lancet Plan Health* 2(7):301–312.

Bowe, B., Xie, Y., Yan, Y. & Al-Aly, Z. 2019. Burden of cause-specific mortality associated with $PM_{2.5}$ air pollution in the United States. *JAMA Netw Open* 2(11): e1915834.

Brunekreef, B. & Holgate, S.T. 2002. Air pollution and health. *Lancet* 360(9341): 1233–1242.

Burnett, R.T., Arden Pope, C., Ezzati, M. et al. 2014. An integrated risk function for estimating the global burden of disease attributable to ambient fine particulate matter exposure. *Environ Health Perspect* 122(4): 397–403.

Castner, J., Guo, L. & Yin, Y. 2018. Ambient air pollution and emergency department visits for asthma in Erie County, New York 2007–2012. *Int Arch Occup Environ Health* 91(2): 205–214.

Cepeda, M., Schoufour, J., Freak-Poli, R., Koolhaas, C.M., Dhana, K., Bramer, W.M. & Franco, O.H. 2017. Levels of ambient air pollution according to mode of transport: A systematic review. *Lancet Public Health* 2(1): e23–e34.

Chambers, S.D. & Podstawczyńska, A., 2019. Improved method for characterising temporal variability in urban air quality part II: Particulate matter and precursors in central Poland, *Atmos Environ* 219: 117040.

Chang, H.H., Pan, A., Lary, D.J. et al. 2019. Time-series analysis of satellite-derived fine particulate matter pollution and asthma morbidity in Jackson, MS. *Environ Monit Assess* 191: 280.

Chen, D., Liu, X., Lang, J. et al., 2017a. Estimating the contribution of regional transport to $PM_{2.5}$ air pollution in a rural area on the North China Plain. *Sci Total Environ* 583: 280–291.

Chen, D., Zhang, F., Yu, C. et al. 2019. Hourly associations between exposure to ambient particulate matter and emergency department visits in an urban population of Shenzhen, China. *Atmos Environ* 209: 78–85.

Chen, T.M., Gokhale, J., Shofer, S. & Kuschner, W.G. 2007. Outdoor air pollution: Nitrogen dioxide, sulfur dioxide, and carbon monoxide health effects. *Am J Med Sci* 333: 249–256.

Cheng, Y.H. 2008. Comparison of the TSI model 8520 and GRIMM series 1.108 portable aerosol instrument used to monitor particulate matter in an iron foundry. *J Occup Environ Hygiene* 5(3): 157–168.

Chlopek, Z. & Strzalkowska, K. 2018. Research on the impact of automotive sources on the immission of specific size fractions of particulate matter in a street canyon. *Arc Automot Eng – Archiwum Motoryzacji* 80(2): 19–35.

Choi, W., Ranasinghe, D., DeShazo, J.R., Kim J.-J. & Paulson, S.E. 2018. Where to locate transit stops: Cross-intersection profiles of ultrafine particles and implications for pedestrian exposure. *Environ Pollut* 233: 235–245.

Climate-data.org. 2020. Climate Lublin - climate graph, temperature graph, climate table. https://pl.climate-data.org/europa/polska/lublin-voivodeship/lublin-622/ (Accessed in October 2020).

Ćwiklak, K., Pastuszka, J.S. & Rogula-Kozłowska, W. 2009. Influence of traffic on particulate-matter polycyclic aromatic hydrocarbons in urban atmosphere of Zabrze, Poland. *Pol J Environ Stud* 18 (4): 579–585.

Czarnecka, M., Nidzgorska-Lencewicz, J. & Rawicki, K. 2017. Thermal conditions and air pollution in selected Polish cities during the winter period in 2016/2017. *Sci Rev Eng Environ Sci* 26(4): 437–446.

Czerwiński, J. 2019. Elemental composition of traffic-related dust deposited in Lublin area. (not published).

Dai, X., Liu, H., Chen, D. & Zhang, J. 2018. Association between ambient particulate matter concentrations and hospitalization for ischemic heart disease (I20-I25, ICD-10) in China: a multicity case-crossover study. *Atmos Environ* 186: 129–35.

Dall'Osto, M., Thorpe, A., Beddows, D.C.S., Harrison, R.M., Barlow, J.F., Dunbar, T., Williams, P.I. & Coe, H. 2010. Remarkable dynamics of nanoparticles in the urban atmosphere. *Atmos Chem Phys Discus* 10(12): 30651–30689.

De Nazelle, A., Bode, O. & Orjuela, J.P. 2017. Comparison of air pollution exposures in active vs. passive travel modes in European cities: A quantitative review. *Environ Int* 99: 151–160.

Dehghan, A., Khanjani, N., Bahrampour, A. et al. 2018. The relation between air pollution and respiratory deaths in Tehran, Iran- using generalized additive models. *BMC Pulm Med* 18(1): 49.

Directive 2002/51/EC 2002 of the European Parliament and of the Council of 19 July on the Reduction of the Level of Pollutant Emissions from Two- and Three-Wheel Motor Vehicles and Amending Directive 97/24/EC (Text with EEA Relevance) - Statement by the C. 2002.

Directive 91/441/EEC 1991 Amending Directive 70/220/EEC on the Approximation of the Laws of the Member States Relating to Measures to Be Taken against Air Pollution by Emissions from Motor Vehicles. 1991.

Directive 98/69/EC 1998 of the European Parliament and of the Council of 13 October 1998 relating to measures to be taken against air pollution by emissions from motor vehicles and amending Council Directive 70/220/EEC.1998.

Donaldson, K. & Beswick, P.S. 1996. Free radical activity associated with the surface of the particles: a unifying factor in deter-mining biological activity. *Toxicol Lett* 88: 293–298.

Dons, E., Int Panis, L., Van Poppel, M., Theunis, J. & Wets, G. 2012. Personal exposure to Black Carbon in transport microenvironments. *Atmos Environ* 55: 392–398.

Du, Y., Xu, X., Chu, M., Guo, Y. & Wang, J. 2016. Air particulate matter and cardiovascular disease: the epidemiological, biomedical and clinical evidence. J Thor Dis 8(1): E8–E19.

Ebersviller, S., Lichtveld, K., Sexton, K.G. et al. 2012. Gaseous VOCs rapidly modify particulate matter and its biological effects – Part 1: simple VOCs and model PM. *Atmos Chem Phys Discuss* 12: 5065–5105.

EC 715/2007. 2007. Regulation (EC) No 715/2007 of the European Parliament and of the Council of 20 June 2007 on type approval of motor vehicles with respect to emissions from light passenger and commercial vehicles (Euro 5 and Euro 6) and on access to vehicle repair and maintenance information.

EC. 2008. Directive 2008/50/EC of the European Parliament and of the Council of 21 May 2008 on ambient air quality and cleaner air for Europe, L 152, 11.06.2008.

EC. 2016. Directive (EU) 2016/2284 of the European Parliament and of the Council of 14 December 2016 on the reduction of national emissions of certain atmospheric pollutants, amending Directive 2003/35/EC and repealing Directive 2001/81/EC, L 344, 17.12.2016.

EEA (European Environmental Agency). 2015. The European environment – state and outlook 2015. Copenhagen.

EEA. 2017. European Environment Agency. Air quality in Europe – 2017 report, http://www.eea.europa.eu/publications/air-quality-in-europe-2017 (Accessed in October 2020).

EEA. 2020a. European Environment Agency (2020). Air quality in Europe – 2020 report, (https://www.eea.europa.eu/publications/air-quality-in-europe-2020-report (Accessed in January 2021).

EEA. 2020b. European Environment Agency (2020). National emission reduction commitments Directive reporting status 2020, (https://www.eea.europa.eu/publications/national-emission-reduction-commitments-directive (Accessed in January 2021).

EMEP/EEA. 2019. EMEP/EEA air pollutant emission inventory guidebook 2019, (https://www.eea.europa.eu/publications/emep-eea-guidebook-2019 (Accessed in October 2020)

Esposito, S., Galeone, C., Lelii, M. et al. 2014. Impact of air pollution on respiratory diseases in children with recurrent wheezing or asthma. *BMC Pulmonary Medicine* 14: 130.

EU 459/2012. 2012. Commission Regulation (EU) No 459/2012 of 29 May 2012 amending Regulation (EC) No 715/2007 of the European Parliament and of the Council and Commission Regulation (EC) No 692/2008 as regards emissions from light passenger and commercial vehicles (Euro 6).

EU 2016/646. 2016. Commission Regulation (EU) 2016/646 of 20 April 2016 amending Regulation (EC) No 692/2008 as regards emissions from light passenger and commercial vehicles (Euro 6).

Eurostat. 2019. Stock of vehicles at regional level - European Commission (Accessed in October 2020).

Eurostat. 2020. Database (Accessed in December 2020).

Fang, X., Fang, B., Wang, C. et al. 2017. Relationship between fine particulate matter, weather condition and daily non-accidental mortality in Shanghai, China: A Bayesian approach. *PLoS One* 12(11): e0187933.

Farhat, A., Mohammadzadeh, A., Balali-Mood, M. et al. 2013. Correlation of blood lead level in mothers and exclusively breastfed infants: a study on infants aged less than six months. *Asia Pac J Med Toxicol* 2: 150–152.

Feng, C., Li, J., Sun, W. et al. 2016. Impact of ambient fine particulate matter ($PM_{2.5}$) exposure on the risk of influenza-like-illness: A time-series analysis in Beijing, China. *Environ Health* 15: 17.

Fruin, S., Westerdahl, D., Sax, T., Sioutas, C. & Fine, P.M. 2008. Measurements and predictors of on-road ultrafine particle concentrations and associated pollutants in Los Angeles. *Atmos Environ* 42: 207–219.

Gandini, M., Scarinzi, C., Bande, S. et al. 2018. Long term effect of air pollution on incident hospital admissions: results from the Italian Longitudinal Study within LIFE MED HISS project. *Environ Int* 121(2): 1087–1097.

Gao, M., Beig, G., Song, S. et al. 2018. The impact of power generation emissions on ambient $PM_{2.5}$ pollution and human health in China and India. *Environ Int* 121(Pt 1): 250–259.

GIOS (Chief Inspectorate for Environmental Protection). 2017. The state of the environment in Poland. Signals 2016 (in Polish). Warszawa, (https://www.gios.gov.pl (Accessed in October 2020).

GIOS (Chief Inspectorate for Environmental Protection). 2020. Air Quality Assessment.

Gładka, A. & Zatoński, T. 2016. Wpływ zanieczyszczenia powietrza na choroby układu oddechowego (in Polish). Kosmos – problemy nauk biologicznych 65(4): 573–582.

Goel, A. & Kumar, P. 2015. Characterisation of nanoparticle emissions and exposure at traffic intersections through fast-response mobile and sequential measurements. *Atmos Environ* 107: 374–390.

Gogna, P., Narain, T.A., O'Sullivan, D.E. et al. 2019. Estimates of the current and future burden of lung cancer attributable to $PM_{2.5}$ in Canada. *Prev Med* 122: 91–99.

Goyer, R.A. 1990. Transplacental transport of lead. *Environ Health Perspect* 89: 101–105.

Greenwell, L.L., Moreno, T., Jones, T.P. et al. 2002. Particle-induced oxidative damage is ameliorated by pulmonary antioxidants. *Free Radic Biol Med* 32: 898–905.

Grynkiewicz-Bylina, B., Rakwic, B. & Pastuszka, J.S. 2005. Assessment of Exposure to Traffic-Related Aerosol and to Particle-Associated PAHs in Gliwice, Poland. *Polish J Environ Stud* 14(1): 117–123.

Guo, H., Kota, S.H., Chen, K. et al. 2018. Source contributions and potential reductions to health effects of particulate matter in India. *Atmos Chem Phys* 18(20): 15219–15229.

He, T., Yang, Z., Liu, T. et al. 2016. Ambient air pollution and years of life lost in Ningbo, China. *Sci Rep* 6: 22485.

Hesterberg, T.W., Bunn, W.B., McClellan, R.O. et al. 2009. Critical review of the human data on short-term nitrogen dioxide (NO_2) exposures: Evidence for NO_2 no-effect levels. *Crit Rev Toxicol* 39: 743–81.

Hinds, W.C. 1999. *Aerosol Technology: Properties, Behaviour and Measurement of Airborne Particles*. New York: John Wiley & Sons Inc.

Hoffmann, B., Weinmayr, G., Hennig, F. et al. 2015. Luftqualität, Schlaganfall und koronare Ereignisse: Ergebnisse der Heinz Nixdorf Recall Studie aus dem Ruhrgebiet. *Dtsch Arztebl Int* 112(12): 195–201.

Holnicki, P., Kałuszko, A., Nahorski, Z., Stankiewicz, K. & Trapp, W. 2017. Air quality modeling for Warsaw agglomeration. Arch Environ Prot 43(1): 48–64.

Hu, X.-M., Nielsen-Gammon, J.W. & Zhang, F. 2010. Evaluation of three planetary boundary layer schemes in the WRF model. *J Appl Meteorol Climatol* 49: 1831–1844.

Huang, K., Yang, X., Liang, F. et al. 2019. Long-Term Exposure to Fine Particulate Matter and Hypertension Incidence in China. *Hypertension* 73(6): 1195–1201.

Huang, L., Zhou, L., Chen, J. et al. 2016. Acute effects of air pollution on influenza-like illness in Nanjing, China: a population-based study. *Chemosphere* 147: 180–187.

Huang, W., Wang, L., Li, J. et al. 2018. Short-term blood pressure responses to ambient fine particulate matter exposures at the extremes of global air pollution concentrations. *Am J Hypertens* 31(5): 590–599.

Husar, R.B. & Renard, W.P. 1998. *Ozone as a Function of Local Wind Speed and Direction: Evidence of Local and Regional Transport*. Pittsburgh, PA: Air and Waste Management Association.

Hvidtfeldt, U.A., Sørensen, M., Geels, C. et al. 2019. Long-term residential exposure to $PM_{2.5}$, PM_{10}, black carbon, NO_2, and ozone and mortality in a Danish cohort. *Environ Int* 123: 265–272.

ICRP. 1994a. *Annals of the International Commission on Radiological Protection ICRP Publication 66: Human Respiratory Tract Model for Radiological Protection; International Commission on Radiological Protection*. Ottawa, ON: Pergamon.

ICRP. 1994b. Human respiratory tract model for radiological protection. A report of a Task Group of the International Commission on Radiological Protection. *Annals of the ICRP* 24(1–3), 1–482.

Janssen, N.A., Fischer, P., Marra, M. et al. 2013. Short-term effects of $PM_{2.5}$, PM_{10} and $PM_{2.5-10}$ on daily mortality in the Netherlands. *Sci Total Environ* 463–464: 20–26.

Janssen, N.A., Hoek, G., Simic-Lawson, M., Fischer, P., Van Bree, L., Ten Brink, H., Keuken, M., Atkinson, R.W., Anderson, H.R. & Brunekreef, B. 2011. Black carbon as an additional indicator of the adverse health effects of airborne particles compared with PM_{10} and $PM_{2.5}$. *Environ Health Perspect* 119(12): 1691.

Jędruszkiewicz, J., Czernecki, B. & Marosz, M. 2017. The variability of PM_{10} and $PM_{2.5}$ concentrations in selected Polish agglomerations: the role of meteorological conditions 2006–2016. *Intern J Environ Health Res* 27(6): 441–462.

Johnson, E. 2003. LPG: a secure, cleaner transport fuel? A policy recommendation for Europe. *Energy Policy* 31(15): 1573–1577.

Johnston, F.H., Salimi, F., Williamson, G.J. et al. 2019. Ambient particulate matter and paramedic assessments of acute diabetic, cardiovascular, and respiratory conditions. *Epidemiology* 30(1): 11–19.

Joodatnia, P., Kumar, P. & Robins, A. 2013. The behaviour of traffic produced nanoparticles in a car cabin and resulting exposure rates. *Atmos Environ* 65: 40–51.

Journal of Laws 2012 Pos. 1031. 2012. Regulation of the Minister of the Environment of 24 August 2012 on the Levels of Certain Substances in the Air. Poland. http://isap.sejm.gov.pl/ (Accessed in October 2020).

Journal of Laws 2019 Pos. 1931. 2019. Regulation of the Minister of Environment of October 8, 2019 Amending the Ordinance on the Levels of Certain Substances in the Air. Poland. http://isap.sejm.gov.pl/ (Accessed in October 2020).

Karagulian, F., Belis, C.A., Dora, C.F., Prüss-Ustün, A.M., Bonjour, S., Adair-Rohani, H. & Amann, M. 2015. Contributions to cities' ambient particulate matter (PM): A systematic review of local source contributions at global level. *Atmos Environ* 120: 475–483.

Karjalainen, P., Pirjola, L., Heikkilä, J., Lähde, T., Tzamkiozis, T., Ntziachristos, L., Keskinen, J. & Rönkkö, T. 2014. Exhaust particles of modern gasoline vehicles: A laboratory and an on-road study. Atmos Environ 97: 262–270.

Kaur, S., Nieuwenhuijsen, M. & Colvile, R. 2005. Personal exposure of street canyon intersection users to $PM_{2.5}$, ultrafine particle counts and carbon monoxide in Central London, UK. *Atmos Environ* 39: 3629–3641.

Kelishadi, R. & Poursafa, P. 2010. Air pollution and non-respiratory health hazards for children. *Arch Med Sci* 6: 483–495.

Kelly, F.J. 2003. Oxidative stress: its role in air pollution and adverse health effects. *Occup Environ Med* 60: 612–616.

Ketzel, M., Omstedt, G., Johansson, C., Düring, I., Pohjola, M., Oettl, D., Gidhagen, L., Wåhlin, P., Lohmeyer, A., Haakana, M. & Berkowicz, R. 2007. Estimation and validation of PM 2.5/PM 10 exhaust and non-exhaust emission factors for practical street pollution modelling. Atmos Environ 41: 9370–9385.

Kim, J.H., Oh, I.H., Park, J.H. & Cheong, H.K. 2018. Premature deaths attributable to long-term exposure to ambient fine particulate matter in the Republic of Korea. *J Korean Med Sci* 33(37): e251.

Kim, Y.K., Jung, J.S., Lee, S.H. et al. 1997. Effects of antioxidants and Ca^{2+} in cisplatin-induced cell injury in rabbit renal cortical slices. *Toxicol Appl Pharmacol* 146: 261–269.

Klejnowski, K., Kozielska, B., Krasa, A. & Rogula-Kozłowska, W. 2010. Polycyclic aromatic hydrocarbons in PM_1, $PM_{2.5}$, PM_{10} and TSP in the Upper Silesian agglomeration, Poland. *Arch Environ Protect* 36(2): 65–72.

Knibbs, L.D., Cole-Hunter, T. & Morawska, L. 2011. A review of commuter exposure to ultrafine particles and its health effects. *Atmos Environ* 45: 2611–2622.

KOBiZE. 2020a. Krajowy Ośrodek Bilansowania i Zarządzania Emisjami (KOBiZE), Instytut Ochrony Środowiska – Państwowy Instytut Badawczy (2020), Krajowy bilans emisji SO2, NOX, CO, NH3, NMLZO, pyłów, metali ciężkich i TZO za lata 1990−2018. Raport syntetyczny (in Polish), (http://www.kobize.pl (Accessed in October 2020).

KOBiZE. 2020b. National Centre for Emissions Management (KOBiZE) at the Institute of Environmental Protection – National Research Institute (2020), Poland's Informative Inventory Report 2020, (http://www.kobize.pl (Accessed in October 2020).

Kondracki, J. 2002. *Geografia regionalna Polski*. Warszawa: PWN.

Konduracka, E., Niewiara, Ł., Guzik, B. et al. 2019. Effect of short-term fluctuations in outdoor air pollution on the number of hospital admissions due to acute myocardial infarction among inhabitants of Kraków, Poland. *Pol Arch Intern Med* 129(2): 884–892.

Kozawa, K.H., Winer, A.M. & Fruin, S.A. 2012. Ultrafine particle size distributions near freeways: Effects of differing wind directions on exposure. *Atmos Environ* 63: 250–260.

Kozielska, B., Rogula-Kozłowska, W. & Pastuszka, J.S. 2013. Traffic emission effects on ambient air pollution by $PM_{2.5}$-related PAH in Upper Silesia, Poland. *Inter J Environ Pollut* 53(3–4): 245–264.

Kuhlbusch, T., John, A., Fissan, H., Schmidt, K.-G., Schmidt, F., Pfeffer, H.-U. & Gladtke, D. 1998. Diurnal variations of particle number concentrations - influencing factors and possible implications for climate and epidemiological studies. *J Aerosol Sci* 29: 213–214.

Kumar, P., Fennell, P. & Britter, R. 2008. Effect of wind direction and speed on the dispersion of nucleation and accumulation mode particles in an urban street canyon. *Sci Total Environ* 402: 82–94.

Kumar, P., Fennell, P., Hayhurst, A. & Britter, R.E. 2009. Street versus rooftop level concentrations of fine particles in a Cambridge street canyon. *Boundary Layer Meteorol* 131: 3–18.

Kumar, A., Singh, B.P., Punia, M. et al. 2014. Assessment of indoor air concentrations of VOCs and their associated health risks in the library of Jawaharlal Nehru University, New Delhi. *Environ Sci Pollut Res Int* 21: 2240–2248.

Kumar, P. & Goel, A. 2016. Concentration dynamics of coarse and fine particulate matter at and around the signalised traffic intersections. *Environ Sci: Processes & Impacts* 18: 1220–1235.

Kumar, P., Morawska, L., Birmili, W., Paasonen, P., Hu, M., Kulmala, M., Harrison, R.M., Norford, L. & Britter, R. 2014. Ultrafine particles in cities. *Environ Inter* 66: 1–10.

Kwak, J.H., Kim, H., Lee, J. & Lee, S. 2013. Characterization of non-exhaust coarse and fine particles from on-road driving and laboratory measurements. *Sci Total Environ* 458–460: 273–282.

Kwak, K.-H., Woo, S.H., Kim, K.H., Lee, S.-B., Bae, G.-N., Ma, Y.-I., Sunwoo, Y. & Baik, J.-J. 2018. On-road air quality associated with traffic composition and street-canyon ventilation: mobile monitoring and CFD modeling. *Atmosphere* 9(3): 92.

Lanzinger, S., Schneider, A. & Breitner, S. et al. 2016. Associations between ultrafine and fine particles and mortality in five central European cities—results from the UFIREG study. *Environ Int* 88: 44–52.

Lee, B.J., Kim, B. & Lee, K. 2014. Air pollution exposure and cardiovascular disease. *Toxicol Res* 30(2), 71–75.

Lee, S., Lee, W., Kim, D., Kim, E., Myung, W., Kim, S.Y. & Kim, H. 2019. Short-term $PM_{2.5}$ exposure and emergency hospital admissions for mental disease. *Environ Res* 171: 313–320.

Leepe, K.A., Li, M., Fang, X. et al. 2019. Acute effect of daily fine particulate matter pollution on cerebrovascular mortality in Shanghai, China: A population-based time series study. *Environ Sci Pollut Res* 26(25): 25491–25499.

Li, D., Wang, J., Zb, Y. et al. 2019. Air pollution exposures and blood pressure variation in type-2 diabetes mellitus patients: A retrospective cohort study in China. *Ecotoxicol Environ Saf* 171: 206–210.

Li, G., Xue, M., Zeng, Q. et al. 2017. Association between fine ambient particulate matter and daily total mortality: an analysis from 160 communities of China. *Sci Total Environ* 599–600: 108–113.

Li, J., Liu, H., Lv, Z. et al. 2018. Estimation of $PM_{2.5}$ mortality burden in China with new exposure estimation and local concentration-response function. *Environ Pollut* 243: 1710–1718.

Li, L., Li, Q., Huang, L. et al., 2020. Air quality changes during the COVID-19 lockdown over the Yangtze River Delta Region: an insight into the impact of human activity pattern changes on air pollution variation. *Sci Total Environ* 732: 139282.

Li, Y., Ma, Z., Zheng, C. & Shang, Y. 2015. Ambient temperature enhanced acute cardiovascular-respiratory mortality effects of $PM_{2.5}$ in Beijing, China. *Int J Biometeorol* 59(12): 1761–1770.

Liang, F., Yang, X., Liu, F. et al. 2019. Long-term exposure to ambient fine particulate matter and incidence of diabetes in China: a cohort study. *Environ Int* 126: 568–575.

Lin, H., Liu, T., Xiao, J. et al. 2016. Quantifying short-term and long-term health benefits of attaining ambient fine particulate pollution standards in Guangzhou, China. *Atmos Environ* 137: 38–44.

Ljungman, P.L.S., Andersson, N., Stockfelt, L. et al. 2019. Long-term exposure to particulate air pollution, black carbon, and their source components in relation to ischemic heart disease and stroke. *Environ Health Perspect* 127(10): 107012.

López-Villarrubia, E., Iñiguez, C., Costa, O. & Ballester, F. 2016. Acute effects of urban air pollution on respiratory emergency hospital admissions in the Canary Islands. *Air Qual Atmos Health* 9(7): 713–722.

Lublin City Office. 2017. Standard of living in Lublin. https://lublin.eu/en/urzad-miasta/ (Accessed in December 2019),

Luengo Oroz, J. & Reis, S. 2019. Assessment of cyclists' exposure to ultrafine particles along alternative commuting routes in Edinburgh. *Atmos Pollut Res* 10(4): 1148–1158.

Makar, M., Antonelli, J., Di, Q. et al. 2017. Estimating the causal effect of low levels of fine particulate matter on hospitalization. *Epidemiology* 28(5): 627–634.

Marczak, H. 2017. Particulate matter in atmospheric air in urban agglomeration. *J Ecol Eng* 18(3): 149–155.

Maricq, M., Podsiadlik, D., Brehob, D. & Haghgooie, M. 1999. Particulate emissions from a direct-injection spark-ignition (DISI) engine. SAE Tech Pap 1999-01-1530.

McGuinn, L.A., Ward-Caviness, C.K., Neas, L.M. et al. 2016. Association between satellite-based estimates of long-term $PM_{2.5}$ exposure and coronary artery disease. *Environ Res* 145: 9–17.

Mehta, M., Chen, L.C., Gordon, T. et al. 2008. Particulate matter inhibits DNA repair and enhances mutagenesis. *Mutat Res* 657: 116–121.

Mejía, J.F., Morawska, L. & Mengersen, K. 2008. Spatial variation in particle number size distributions in a large metropolitan area. *Atmos Chem Phys* 8: 1127–1138.

Meng, X., Wu, Y., Pan, Z., Wang, H., Yin, G. & Zhao, H. 2019. Seasonal characteristics and particle-size distributions of particulate air pollutants in Urumqi. *Int J Environ Res Public Health* 16(3): 396.

Molhave, L., Clausen, G., Berglund, B. et al. 2004. Total Volatile Organic Compounds (TVOC) in Indoor Air Quality Investigations. *Indoor Air* 7: 225–240.

Mönkkönen, P., Koponen, I.K., Lehtinen, K.E.J., Hämeri, K., Uma, R. & Kulmala, M. 2005. Measurements in a highly polluted Asian mega city: observations of aerosol number size distribution, modal parameters and nucleation events. *Atmos Chem Phys* 5: 57–66.

Moreno, T., Pacitto, A., Fernández, A., Amato, F., Marco, E., Grimalt, J.O., Buonanno, G. & Querol, X. 2019. Vehicle interior air quality conditions when travelling by taxi. *Environ Res* 172: 529–542.

Moreno, T., Reche, C., Rivas, I., Minguillón, M.C., Martins, V., Vargas, C., Buonanno, G., Parga, J., Pandolfi, M., Brines, M., Ealo, M., Fonseca, A.S., Amato, F., Sosa, G., Capdevila, M., de Miguel, E., Querol, X. & Gibbons, W. 2015. Urban air quality comparison for bus, tram, subway and pedestrian commutes in Barcelona. *Environ Res* 142: 495–510.

Moulton, P.V. & Yang, W. 2012. Air pollution, oxidative stress, and Alzheimer's disease. J Environ Public Health 25: 472751.

Mukerjee, S., Smith, L., Brantley, H., Stallings, C., Neas, L., Kimbrough, S. & Williams, R. 2015. Comparison of modeled traffic exposure zones using on–road air pollution measurements. *Atmos Pollut Res* 6(1): 82–87.

Myung, W., Lee, H. & Kim, H. 2019. Short-term air pollution exposure and emergency department visits for amyotrophic lateral sclerosis: A time-stratified case-crossover analysis. *Environ Int* 123: 467–475.

Nascimento, L.F.C., Vieira, L.C.P.F., Mantovani, K.C.C. & Moreira, D.S. 2016. Poluição do ar e doenças respiratórias: Estudo ecológico de série temporal. *Sao Paulo Med J* 134(4): 315–321.

Nave, R. & Mueller, H. 2013. From inhaler to lung: clinical implications of the formulations of ciclesonide and other inhaled corticosteroids. Int. J. Gen. Med. 6: 99–107.

Nhung, N.T.T., Schindler, C., Dien, T.M. et al. 2018. Acute effects of ambient air pollution on lower respiratory infections in Hanoi children: an eight-year time series study. *Environ Int* 110: 139–148.

NIH (National Institute of Environmental Health Sciences). 2013. Lead and Your Health. https://www.niehs.nih.gov/health/materials/lead_and_your_health_508.pdf (Accessed 2020).

Nzwalo, H., Guilherme, P., Nogueira, J. et al. 2019. Fine particulate air pollution and occurrence of spontaneous intracerebral hemorrhage in an area of low air pollution. *Clin Neurol Neurosurg* 176: 67–72.

Oleniacz, R., Bogacki, M., Szulecka, A., Rzeszutek, M. & Mazur, M. 2016. Assessing the impact of wind speed and mixing-layer height on air quality in Krakow (Poland) in the years 2014–2015. *J Civil Eng, Environ Architect* 63: 315–342.

Olszowski, T. 2016. Changes in PM_{10} concentration due to large-scale rainfall. *Arab J Geosci* 9: 160.

Paasonen, P., Visshedjik, A., Kupiainen, K., Klimont, Z., Denier van der Gon, H. & Kulmala, M. 2013. Aerosol particle number emissions and size distributions: Implementation in the GAINS model and initial results. IIASA Interim Report. IIASA, IR-13-020.

Park, G., Mun, S., Hong, H. et al. 2019. Characterization of emission factors concerning gasoline, LPG, and diesel vehicles via transient chassis-dynamometer tests. *Appl Sci* 9: 1573.

Pawłowski, L., Kozak, Z., Łyżwa, J., Wyszkowski, A., Adamczyk, M., Mihułka, A. & Pazdan, R. Motoryzacyjne skażenia środowiska miasta Lublina w 1995 r., Lublin 1995.

Penkała, M., Ogrodnik, P. & Rogula-Kozłowska, W. 2018. Particulate matter from the road surface abrasion as a problem of non-exhaust emission control. *Environments* 5(1): 9.

Pinault, L., Tjepkema, M. Crouse, D.L. et al. 2016. Risk estimates of mortality attributed to low concentrations of ambient fine particulate matter in the Canadian community health survey cohort. *Environ Health* 15: 18.

Pinault, L.L., Weichenthal, S. Crouse, D.L. et al. 2017. Associations between fine particulate matter and mortality in the 2001 Canadian census health and environment cohort. *Environ Res* 159: 406–415.

Piotrowicz, A. & Polednik, B. 2019. Exposure to aerosols particles on an urban road. *J Ecol Eng* 20(5): 27–34.

Pirjola, L., Lähde, T., Niemi, J.V., Kousa, A., Rönkkö, T., Karjalainen, P., Keskinen, J., Frey, A. & Hillamo, R. 2012. Spatial and temporal characterization of traffic emissions in urban microenvironments with a mobile laboratory. *Atmos Environ* 63: 156–167.

Polednik, B. & Piotrowicz, A. 2020. Pedestrian exposure to traffic-related particles along a city road in Lublin, Poland. *Atmos Pollut Res* 11(4): 686–692.

Polednik, B. 2013a. Variations in particle concentrations and indoor air parameters in classrooms in the heating and summer season. *Arch Environ Protect* 39: 15–28.

Polednik, B. 2013b. Particulate matter and student exposure in school classrooms in Lublin, Poland. *Environ Res* 120: 134–139.

Polednik, B. 2013c. Particle exposure in a baroque church during Sunday masses, *Environ Res* 126: 215–220.

Polednik, B., Piotrowicz, A., Pawłowski, L. & Guz, Ł. 2018. Traffic-related particle emissions and exposure on an urban road. *Archives of Environmental Protection* 2(44): 83–93.

Polednik, B. 2021. Exposure of staff to aerosols and bioaerosols in a dental office. Build Environ 185: 1–13.

Polichetti, G., Capone, D., Grigoropoulos, K. et al. 2013. Effects of ambient air pollution on birth outcomes: an overview. *Crit Rev Environ Sci Technol* 43(7): 752–774.

Polichetti, G., Cocco, S., Spinali, A. et al. 2009. Effects of particulate matter (PM(10), PM (2.5) and PM (1)) on the cardiovascular system. *Toxicology* 261(1–2): 1–8.

Pope, C.A., Ezzati, M. & Dockery, D.W. 2009. Fine particulate air pollution and life expectancy in the United States. New Eng J Med 360(4): 376–386.

Pope, C.A., Lefler, J.S., Ezzati, M. et al. 2019. Mortality risk and fine particulate air pollution in a large, representative cohort of U.S. adults. *Environ Health Perspect* 127(7): 77007.

Pothirat, C., Chaiwong, W., Liwsrisakun, C. et al. 2019. The short-term associations of particular matters on non-accidental mortality and causes of death in Chiang Mai, Thailand: a time series analysis study between 2016–2018. *Int J Environ Health Res* 10: 1–10.

Pozzer, A., Bacer, S., Sappadina, S.D.Z. et al. 2019. Long-term concentrations of fine particulate matter and impact on human health in Verona, Italy. *Atmos Pollut Res* 10(3): 731–738.

Prashant, K., Fennell, P.S., Hayhurst, A.N. & Britter. R.E. 2009. Street versus rooftop level concentrations of fine particles in a Cambridge street canyon. *In Boundary-Layer Meteor* 131:3–18.

Puett, R.C., Hart, J.E., Yanosky, J.D., Spiegelman, D., Wong, M., Fisher, J.A., Hong, B. & Laden, F. 2014. Particulate matter air pollution exposure, distance to road, and incident lung cancer in the Nurses' Health Study cohort. *Environ Health Perspect* 122(9): 926–932.

Qiu, Z., Lv, H., Wang, W., Zhang, F., Wang, W. & Hao, Y. 2019a. Pedestrian exposure to PM$_{2.5}$, BC and UFP of adults and teens: A case study in Xi'an, China. *Sustain Cities Soc* 51: 101774.

Qiu, Z., Wang, W., Zheng, J. & Lv, H. 2019b. Exposure assessment of cyclists to UFP and PM on urban routes in Xi'an, China. *Environ Pollut* 250: 241–250.

Qu, F., Liu, F., Zhang, H. et al. 2019. The hospitalization attributable burden of acute exacerbations of chronic obstructive pulmonary disease due to ambient air pollution in Shijiazhuang, Chin. *Environ Sci Pollut Res* 26(30): 30866–30875.

Rahman, I. & Macnee, W. 1996. Role of oxidants/antioxidants in smoking-induced lung diseases. *Free Radic Biol Med* 21: 669–681.

Rakowska, A., Wong, K.C., Townsend, T., Chan, K.L., Westerdahl, D., Ng, S., Močnik, G., Drinovec, L. & Ning, Z. 2014. Impact of traffic volume and composition on the air quality and pedestrian exposure in urban street canyon. *Atmos Environ* 98: 260–270.

Raub, J.A. & Benignus, V.A. 2002. Carbon monoxide and the nervous system. *Neurosci Biobehav Rev* 26(8): 925–940.

RBA. 2018. Road and Bridge Authority in Lublin (in Polish) http://www.zdm.lublin.eu/?page_id=1716 (Accessed in June 2019).

Renzi, M., Cerza, F., Gariazzo, C. et al. 2018. Air pollution and occurrence of type 2 diabetes in a large cohort study. *Environ Int* 112: 68–76.

Requia, W.J., Adams, M.D. & Koutrakis, P. 2017. Association of PM$_{2.5}$ with diabetes, asthma, and high blood pressure incidence in Canada: a spatiotemporal analysis of the impacts of the energy generation and fuel sales. *Sci Total Environ* 584–585: 1077–1083.

Richmont-Bryant, J., Owen, R.C., Graham, S. et al. 2017. Estimation of on-road NO$_2$ concentrations, NO$_2$/NO$_X$ ratios, and related roadway gradients from near-road monitoring data. *Air Qual Atm Health* 10: 611–625.

Rogula-Kozłowska, W., Klejnowski, K., Rogula-Kopiec, P., Ośródka, L., Krajny, E., Błaszczak, B. & Mathews, B. 2014. Spatial and seasonal variability of the mass concentration and chemical composition of PM$_{2.5}$ in Poland. *Air Qual Atmos Health* 7(1): 41–58.

Rogula-Kozłowska, W., Mach, T., Rogula-Kopiec, P., Rybak, J. & Nocoń, K. 2019. Concentration and elemental composition of quasi-ultrafine particles in Upper Silesia. *Environ Prot Eng* 45(1): 171–184.

Rogula-Kozłowska, W., Majewski, G., Błaszczak, B., Klejnowski, K. & Rogula-Kopiec, P. 2016. Origin-oriented elemental profile of fine ambient particulate matter in central European suburban conditions. *Int J Environ Res Public Health* 13(7): 715.

Rogula-Kozłowska, W., Pastuszka, J.S. & Talik, E. 2008. Influence of vehicular traffic on concentration and particle surface composition of PM$_{10}$ and PM$_{2.5}$ in Zabrze, Poland. *Polish Journal of Environmental Studies* 17(4): 539–548.

Rose, J.J., Bocian, K.A., Xu, Q. et al. 2020. A neuroglobin-based high-affinity ligand trap reverses carbon monoxide-induced mitochondrial poisoning. *J Biol Chem* 295: 6357–6371.

Rozbicka, K. & Michalak, M. 2015. Characteristic of selected air pollutants concentration in Warsaw (Poland). *Sci Rev Eng Environ Sci* 24(2): 193–206.

Ruuskanen, J., Tuch, T., Brink, H.T. et al. 2001. Concentrations of ultrafine, fine and $PM_{2.5}$ particles in three European Cities. *Atmos Environ* 35(21): 3729–3738.

Sabaliauskas, K., Jeong, C.-H., Yao, X., Jun, Y.-S., Jadidian, P. & Evans, G.J. 2012. Five-year roadside measurements of ultrafine particles in a major Canadian city. *Atmos Environ* 49: 245–256.

Sahu, S.K., Zhang, H., Guo, H. et al. 2019. Health risk associated with potential source regions of $PM_{2.5}$ in Indian cities. *Air Qual Atmos Health* 12(3): 327–340.

Santos, U.P., Ferreira Braga, A.L., Bueno Garcia, M.L. et al. 2019. Exposure to fine particles increases blood pressure of hypertensive outdoor workers: a panel study. *Environ Res* 174: 88–94.

Sartini, C., Zauli Sajani, S., Ricciardelli, I., Delgado-Saborit, J.M., Scotto, F., Trentini, A., Ferrari, S. & Poluzzi, V. 2013. Ultrafine particle concentrations in the surroundings of an urban area: comparing downwind to upwind conditions using Generalized Additive Models (GAMs). *Environ Sci: Processes & Impacts* 11: 2087–2095.

Shi, X. & Brasseur, G.P. 2020. The response in air quality to the reduction of Chinese economic activities during the COVID-19 outbreak. *Geophys Res Lett* 47(11), e2020GL088070.

Skubacz, K. 2009. Measurements of aerosol size distribution in urban areas of Upper Silesia. *Arch Environ Protect* 35(4): 23–34.

Slama, A., Śliwczyński, A., Woźnica, J. et al. 2019. Impact of air pollution on hospital admissions with a focus on respiratory diseases: a time-series multi-city analysis. *Environ Sci Pollut Res* 26: 16998–17009.

Song, J., Liu, Y., Zheng, L. et al. 2018. Acute effects of air pollution on type II diabetes mellitus hospitalization in Shijiazhuang, China. *Environ Sci Pollut Res* 30: 30151–30159.

Sówka, I. (2017). Road transport as a source of air pollution in urban agglomerations. *Czysta Energia* 1–2: 24–28.

Sówka, I., Chlebowska-Styś, A., Pachurka, Ł. & Rogula-Kozłowska, W. 2018. Seasonal variations of $PM_{2.5}$ and PM_{10} concentrations and inhalation exposure from PM-bound metals (As, Cd, Ni): first studies in Poznań (Poland). *Arch Environ Protect* 44(4): 86–95.

Sówka, I.M., Chlebowska-Styś, A., Pachurka, Ł., Rogula-Kozłowska, W. & Mathews, B. 2019. Analysis of particulate matter concentration variability and origin in selected urban areas in Poland. *Sustainability* 11: 5735.

Sram, R.J., Veleminsky, M.Jr, Veleminsky, M.Sr & Stejskalová, J. 2017. The impact of air pollution to central nervous system in children and adults. *Neuro Endocrinol Lett* 38(6): 389–396.

Srimuruganandam, B. & Shiva Nagendra, S.M., 2010. Analysis and interpretation of particulate matter–PM_{10}, $PM_{2.5}$ and $PM_{1.0}$ emissions from the heterogeneous traffic near an urban roadway. *Atmos Pollut Res* 1: 184–194.

Stanek, L.W., Sacks, J.D., Dutton, S.J. & Dubois, J.J.B. 2011. Attributing health effects to apportioned components and sources of particulate matter: An evaluation of collective results. *Atm Environ* 45: 5655–5663.

Stanier, C.O., Khlystov, A.Y. & Pandis, S.N. 2004. Nucleation events during the Pittsburgh air quality study: Description and relation to key meteorological, gas phase, and aerosol parameters. *Aerosol Sci Technol* 38(1): 253–264.

Statistics Poland. 2013. Area and population in the territorial profile in 2013. https://stat.gov.pl/obszary-tematyczne/ludnosc/ludnosc/powierzchnia-i-ludnosc-w-przekroju-terytorialnym-w-2013-r-,7,10.html (Accessed in December 2019).

Statistics Poland. 2017. Lublin in figures 2017 (Accessed in December 2019).

Statistics Poland. 2019. Current research results - Demographics and demographics database. stat.gov.pl (Accessed in December 2019).

Statistics Poland. 2020a. Data for Territorial Division Unit. https://bdl.stat.gov.pl/BDL/start (Accessed in December 2020).

Statistics Poland. 2020b. Energy efficiency in Poland in years 2008–2018 (Accessed in December 2020).

Suchorab, Z., Sobczuk, H., Guz, Ł. & Łagód, G. 2017. Gas sensors array as a device to classify mold threat of the buildings. in M. Pawłowska & M. Pawłowski (eds.), *Environmental Engineering*, Taylor and Francis Group.

Szczygłowski, P. & Mazur, M. 2008. Application of BOOT statistical package in calculating pollutant spreading in air, *Environ Protect Eng* 34(4): 151–156.

Taczanowski, J., Kołoś, A., Gwosdz, K., Domański, B. & Guzik, R. 2018. The development of low-emission public urban transport in Poland. Bulletin of Geography. *Socio-economic Series* 41: 79–92.

Tainio, M. 2015. Burden of disease caused by local transport in Warsaw, Poland. *J Trans Health* 2(3): 423–433.

Thurston, G.D., Ahn, J., Cromar, K.R. et al. 2016. Ambient particulate matter air pollution exposure and mortality in the NIH-AARP diet and health cohort. *Environ Health Perspect* 124(4): 484–490.

Tian, Y., Liu, H., Xiang, X. et al. 2019. Ambient coarse particulate matter and hospital admissions for ischemic stroke: a national analysis. *Stroke* 50(4): 813–819.

TomTom. 2019. Traffic index. http://www.tomtom.com/trafficindex (Accessed in July 2020).

US EPA. 2019. Table of Historical SO_2 NAAQS, Sulfur US EPA. https://www3.epa.gov/ttn/naaqs/standards/so2/s_so2_history.html (Accessed in December 2019).

Vahedian, M., Khanjani, N., Mirzaee, M. & Koolivand, A. 2017. Associations of short-term exposure to air pollution with respiratory hospital admissions in Arak, Iran. *J Environ Health Sci Eng* 15(1): 1–16.

Valavanidis, A., Fiotakis, K., Bakeas, E. et al. 2005. Electron paramagnetic resonance study of the generation of reactive oxygen species catalysed by transition metals and quinoid redox cycling by inhalable ambient particulate matter. *Redox Rep* 10: 37–51.

Valavanidis, A., Fiotakis, K. & Vlachogianni, T. 2008. Airborne particulate matter and human health: Toxicological assessment and importance of size and composition of particles for oxidative damage and carcinogenic mechanisms. *J Environ Sci Health* 26: 339–362.

Walton, E.L. 2018. Tainted air: The link between pollution and Alzheimer's disease. Biomed J 41(3): 137–140.

Wang, C., Feng, L. & Chen, K. 2019. The impact of ambient particulate matter on hospital outpatient visits for respiratory and circulatory system disease in an urban Chinese population. *Sci Total Environ* 666: 672–679.

Wang, S., Li, Y., Niu, A. et al. 2018. The impact of outdoor air pollutants on outpatient visits for respiratory diseases during 2012–2016 in Jinan, China. *Respir Res* 19(1): 246.

Warkentin, M.T., Lam, S. & Hung, R.J. 2019. Determinants of impaired lung function and lung cancer prediction among never-smokers in the UK Biobank cohort. *EBioMedicine* 47: 58–64.

Weaver, L.K, Valentine, K.J. & Hopkins, R.O. 2007. Carbon monoxide poisoning: risk factors for cognitive sequelae and the role of hyperbaric oxygen. *Am J Respir Crit Care Med* 176(5): 491–497.

Wehner, B. & A Wiedensohler. 2003. Long term measurements of submicrometer urban aerosols: statistical analysis for correlations with meteorological conditions and trace gases. *Atmos Chem Phys* 3(3): 867–879.

Wheida, A., Nasser, A., El Nazer, M. et al. 2018. Tackling the mortality from long-term exposure to outdoor air pollution in megacities: Lessons from the Greater Cairo case study. *Environ Res* 160: 223–231.

WHO. 2005. Health effects of transport-related air pollution. http://www.euro.who.int/__data/assets/pdf_file/0006/74715/E86650.pdf (Accessed in July 2019).

WHO. 2006. Air Quality Guidelines: Global update 2005- Particulate matter, ozone, nitrogen dioxide and sulfur dioxide. http://www.euro.who.int/en/home (Accessed in December 2018).

WHO. 2007. Health relevance of particulate matter from various sources. Report of a WHO Workshop. World Health Organization, Regional Office for Europe, Copenhagen.

WHO. 2010. Exposure to dioxins and dioxin-like substances: a major public health concern. https://www.who.int/ipcs/features/dioxins.pdf (Accessed in August 2020).

WHO. 2013. WHO Regional Office for Europe. Review of evidence on health aspects of air pollution–REVIHAAP Project. http://www.euro.who.int/__data/assets/pdf_file/0004/193108/REVIHAAP-Final-technical-report-final-version.pdf (Accessed in July 2019).

WHO. 2000. Regional Office of Europe. https://euro.who.int/_data/assets/pdf_file/0020/123086/AQG2ndEd_7_4Sulfuroxide.pdf (Accessed in December 2020).

Wilson, W.E. & Suh, H.H. 1997. Fine particles and coarse particles: concentration relationships relevant to epidemiologic studies. *J Air Waste Manag Assoc* 47: 1238–1249.

WIOS (Voivodship Environmental Protection Inspectorate in Lublin). 2016. Lublin Voivodeship Environment analysis report in years 2013–2015 (in Polish), http://envir.wios.lublin.pl (Accessed in September 2017).

WIOS (Voivodship Environmental Protection Inspectorate in Lublin). 2017. Air Quality Monitoring System (in Polish), http://envir.wios.lublin.pl (Accessed in December 2017).

WIOS (Voivodship Environmental Protection Inspectorate in Lublin). 2018. Report on the state of the environment of lubelskie voivodship in 2017 (in Polish), http://envir.wios.lublin.pl (Accessed in November 2018).

Wolf, K., Stafoggia, M., Cesaroni, G. et al. 2015. Long-term exposure to particulate matter constituents and the incidence of coronary events in 11 European cohorts. *Epidemiology* 26(4): 565–574.

Woo, K.S., Chen, D.R., Pui, D.Y.H. & Mcmurry, P.H. 2001. Measurement of Atlanta Aerosol Size Distributions: Observations of Ultrafine Particle Events. *Aerosol Sci & Technol* 34(1): 75–87.

Wróbel, A., Rokita, E. & Maenhaut, W. 2000. Transport of traffic-related aerosols in urban areas. *Science of the Total Environment* 257(2–3): 199–211.

Wu, Z., Hu, M., Lin, P., Liu, S., Wehner, B. & Wiedensohler, A. 2008. Particle number size distribution in the urban atmosphere of Beijing, China. Atmos Environ 42(34): 7967–7980.

Xing, W.J., Kong, F.J., Li, G.W. et al. 2011. Calcium-sensing receptors induce apoptosis during simulated ischaemia-reperfusion in Buffalo rat liver cells. *Clin Exp Pharmacol Physiol* 38: 605–612.

Xu, A., Mu, Z., Jiang, B. et al. 2017. Acute effects of particulate air pollution on ischemic heart disease hospitalizations in Shanghai, China. *Int J Environ Res Public Health* 14(2): 168.

Xu, K.J., Cui, K.P., Young, L.H. et al., 2020. Impact of the COVID-19 event on air quality in central China. *Aerosol Air Qual Res* 20: 915–929.

Ye, D., Klein, M., Mulholland, J.A. et al. 2018. Estimating acute cardiovascular effects of ambient $PM_{2.5}$ metals. *Environ Health Perspect* 126(2): 027007.

Zalakeviciute, R., López-Villada, J. & Rybarczyk, Y. 2018. Contrasted effects of relative humidity and precipitation on urban $PM_{2.5}$ pollution in high elevation urban areas. *Sustainability* 10(6): 2064.

Zhang, L., Yang, Y., Li, Y. et al. 2019b. Short-term and long-term effects of $PM_{2.5}$ on acute nasopharyngitis in 10 communities of Guangdong, China. *Sci Total Env* 688: 136–142.

Zhang, Z., Chai, P., Wang, J. et al. 2019a. Association of particulate matter air pollution and hospital visits for respiratory diseases: a time-series study from China. *Environ Sci Pollut Res Int* 26(12): 12280–12287.

Zhao, H., Che, H., Ma, Y., Wang, Y., Yang, H., Liu, Y., Wang, Y., Wang, H. & Zhang, X. 2017. The relationship of PM variation with visibility and mixing-layer height under hazy/foggy conditions in the multi-cities of northeast China. *Int J Environ Res Publ Health* 14(5): 471.

Zhao, Y.B., Zhang, K., Xu, X.T. et al. 2020. Substantial changes in nitrogen dioxide and ozone after excluding meteorological impacts during the COVID-19 outbreak in mainland China. *Environ Sci Technol Lett* 7: 402–408.

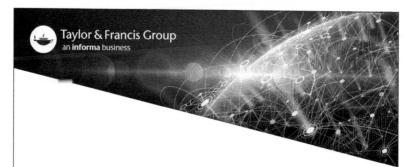

Milton Keynes UK
Ingram Content Group UK Ltd.
UKHW020844141024
449569UK00003B/69